The Institute of Biology's
Studies in Biology no. 6

Microecology

by J. L. Cloudsley-Thompson
M.A., Ph.D., D.Sc., F.L.S., F.I. Biol., F.W.A.
Reader in Zoology, Birkbeck College,
University of London: Formerly
Professor of Zoology, University of Khartoum
and Keeper, Sudan Natural History Museum

Edward Arnold (Publishers) Ltd

First published 1967
by Edward Arnold (Publishers) Ltd,
41 Maddox Street,
London, W1R 0AN

Reprinted 1969
Reprinted 1972

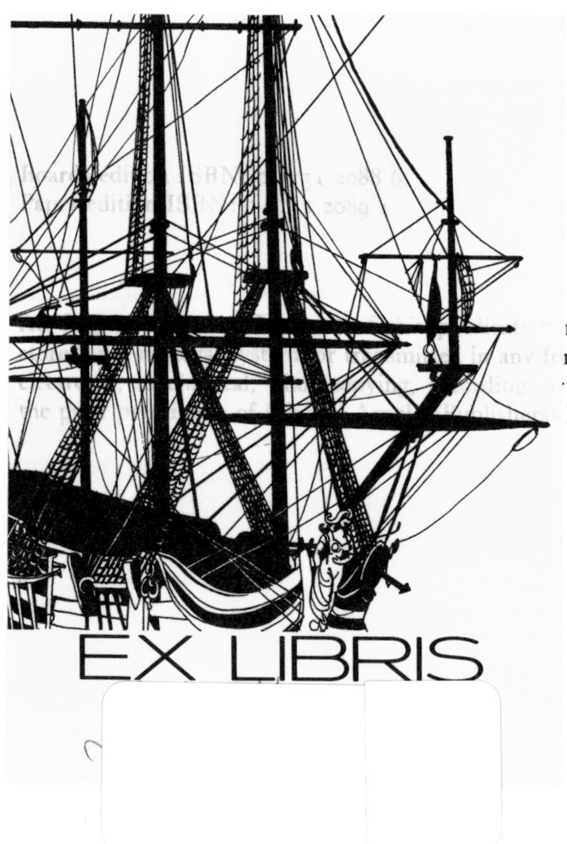

Printed offset in Great Britain by
The Camelot Press Ltd, London and Southampton

General Preface to the Series

It is no longer possible for one textbook to cover the whole field of Biology and to remain sufficiently up to date. At the same time students at school, and indeed those in their first year at universities, must be contemporary in their biological outlook and know where the most important developments are taking place.

The Biological Education Committee, set up jointly by the Royal Society and the Institute of Biology, is sponsoring, therefore, the production of a series of booklets dealing with limited biological topics in which recent progress has been most rapid and important.

A feature of the series is that the booklets indicate as clearly as possible the methods that have been employed in elucidating the problems with which they deal. There are suggestions for practical work for the student which should form a sound scientific basis for his understanding.

1967

INSTITUTE OF BIOLOGY
41 Queen's Gate
London, S.W.7.

Preface

Ecology is one of the more important aspects of biology in the world today. Not only is an ecological approach essential to a satisfactory solution to most problems in applied biology, but the subject is rewarding both as a vehicle for pure research and as a medium of instruction in biological principles. This booklet is intended as an introduction and practical guide to a branch of ecology that can be studied almost anywhere and with minimum requirements of apparatus and laboratory accommodation. I hope that it may introduce the reader to a wealth of interest and pleasure.

Khartoum, Sudan J. L. C.-T.
1967

Contents

Introduction

Ecology is the term used to describe the study of animals and plants in relation to their environment. This includes the physical environment in which they live and their biotic environment—that is, their relationships to other living organisms. It is the modern, scientific approach to natural history but, whilst natural history consists in the observation and description of various types of environment and the animals and plants living in them, ecology also involves the collection and analysis of precise data and the drawing of conclusions from it.

Ecology covers a vast field of biology and the solution to a particular problem may require several different lines of approach. In this booklet, I shall consider only the study of small-scale environments and habitats, such as crevices in soil and leaf-litter, the spaces underneath rocks and stones and beneath the bark of trees and fallen logs, holes in walls, caves, ants' nests and so on. Even this restriction in scope leaves me an enormous subject which has been greatly increased by recent developments and refinements in technique. Many of these are so complicated, however, that they are normally employed only by specialists. I will therefore confine this account to methods that can be used reliably by intelligent students and require neither specialized knowledge nor expensive apparatus. In this I am not adopting a patronizing or parochial attitude: in my own research I always choose the simplest possible methods that will provide the data I require. This is partly to save time and labour, and partly because I tend to mistrust complicated apparatus that may give misleading results. By means of simple observations and experiments, it is often possible to make quite interesting and exciting discoveries.

The accurate naming of plants and animals is an essential preliminary to other biological studies and the beginner often finds himself quite overwhelmed by the number of species with which he is confronted. It is almost impossible, these days, for a single individual to become an expert on more than one or two groups of animals and, even for this, access to a large reference collection is usually necessary. At first sight, therefore, it might appear that the field biologist is in a hopeless situation as he must rely for identification upon the help of his overworked colleagues in national museums. There is, however, a somewhat arbitrary way of dealing with the problem which provides a practical solution for the beginner. This is to ignore rare and difficult species and to concentrate upon those easily recognized in the field. There are two aspects to ecology: *autecology*, the study of individual organisms or species in relation to their habitats, and *synecology*, the study of associations of organisms in relation to an area or habitat. Clearly the autecological approach is likely to be the easier and more productive at first.

Autecological studies of small animal species in relation to their microhabitats have much to commend them. Firstly, they can be most interesting and add greatly to one's enjoyment of nature. Secondly, they are instructive in illustrating general biological principles and, thirdly, they provide a method by which pest species may be studied with a view to their subsequent control. In this booklet I hope to provide an outline of the ways in which such studies, in particular, can be conducted by students so that the results obtained by them are really rewarding. In order that students all over the world may be able to use it, I have not confined the discussion to the microecology of temperate regions only. Although some knowledge of ecological principles is assumed, I have tried to make the account self-explanatory.

Chemistry and physics are very precise subjects compared with the biological sciences and, of these, ecology is probably one of the least precise. This does not imply that one should not try to be as accurate as possible, but it does mean that the ecologist has to be much more than a mere computer. Not only must he obtain statistical correlations, but he has to discover the causal relationships between them. And this often requires insight based on laboratory experiments and an intimate knowledge of the animals and plants with which he is dealing.

Observation and description, as I have said, must be followed by the analysis of data and the drawing of conclusions from it. The fact that a species is found in a particular habitat does not, however, prove that this represents its optimum environment, but merely that it is able to exist there. Indiscriminate predation may enable two species to co-exist when otherwise one would undoubtedly eliminate the other through competition. Again, investigation of gut contents will show what an animal has been feeding on but does not necessarily indicate what it would have eaten if other food had also been available. Examples such as these show why considerable care must be taken in drawing conclusions and formulating hypotheses. It is all too easy to make dreadful mistakes. On the other hand, because of the very nature of the material on which he is working, the ecologist who is over-cautious may never reach any definite conclusions at all. The natural world is so infinitely complex that the best analyses can at present be little more than generalizations or inspired guesses. In this lies its fascination and its challenge.

Microhabitats 1

1.1 Biological principles

A microhabitat may be defined as a self-contained entity which reflects in miniature the biological balance of living organisms in general. The study of microecology is therefore valuable in that it illustrates general biological principles in addition to producing specific problems of its own.

In all ecological studies the physical and chemical environment almost always needs to be investigated, then the organisms, plant and animal that inhabit this environment. These, in turn, can be considered qualitatively—that is, a taxonomic study of the different species present may be made—and quantitatively. The numbers of each species and their interrelationships, one to another, may have to be assessed and this is a laborious task. Nor is the picture thus obtained a true representation of conditions in nature, because diurnal and seasonal changes occur long before equilibrium has been reached, so that a static condition is never attained. Moreover, organisms are continually becoming adapted physiologically and genetically to long-term climatic and geological changes.

1.2 Soils

Soils develop through weathering of the parent rocks that make up the crust of the earth. In addition to this mineral substrate in which the vegetation takes root, soils include dead organic matter and humus, water and air. Weathering of the rocks to produce the *parent material* of the soil is usually achieved by a combination of three processes: (1) *mechanical weathering*, in which breakdown of the rock itself occurs without chemical change; (2) *chemical weathering*, where an actual chemical change of the rock minerals takes place and new substances are formed; and (3) *biological weathering*, where the agents of change are plants and animals.

As the result of such weathering, a characteristic layered arrangement known as a *soil profile* is developed. This depends largely upon the amount and kind of organic matter present and also upon the way in which water falling on the surface removes and redeposits the soluble constituents of the surface layers. Under humid conditions, soluble salts are *leached* away but, where water is scarce, *pedocals*, or soils rich in calcium, are formed. *Pedalfers* or leached soils, occur under humid conditions where rainfall is in excess of potential evaporation.

A soil profile consists of several horizons (Fig. 1–1), each having characteristic physical and chemical properties. At the surface there is frequently a layer of undecomposed material, the litter or *L* layer (often called Fö layer, from the Swedish term, Förna). Beneath this lies the humus or A_0 layer composed of amorphous organic matter which has lost its original structure. Then comes a varying number of A layers of true soil. The first of these, A_1, is a dark-coloured horizon containing a relatively high content of organic matter mixed with mineral fragments. It tends to be thick in savannah and thin in forest soils. The A_2 horizon is frequently ashy grey and is the zone of maximum leaching. The underlying B horizons tend to be darker in colour because they are enriched by iron compounds, clay and humus. A lighter coloured C horizon of parent material then grades into the D horizon of bed rock.

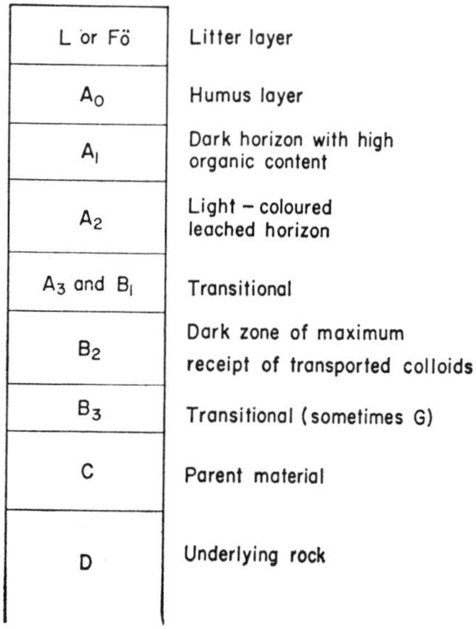

L or Fö	Litter layer
A_0	Humus layer
A_1	Dark horizon with high organic content
A_2	Light – coloured leached horizon
A_3 and B_1	Transitional
B_2	Dark zone of maximum receipt of transported colloids
B_3	Transitional (sometimes G)
C	Parent material
D	Underlying rock

Fig. 1–1 Nomenclature of soil horizons.

This description is applicable to *podsols* (Fig. 1–2), acid soils with strongly acid hydrogen ion concentration (usually below pH 5·5) and excessive drainage. Podsols are developed on sandstones under conditions of moderately heavy rainfall. In *brown earths* or *brown forest earths*, the profile is less uniformly coloured throughout, with a darker humus-rich A_1 horizon on top which grades into slightly lighter coloured subsoil. Brown earths are usually slightly acid and certainly never base-saturated. They are frequently developed over clays and drainage is often impeded to some extent.

Low pH results in an accumulation of litter in which bacteria are comparatively inactive and in which fungi, while they break down cellulose, have relatively little action on lignin. The *raw humus* that results is sometimes called *mor* and is often the product of coniferous trees.

On alkaline rock such as chalk and carboniferous limestone are developed *rendzinas* (Fig. 1–2) or 'humus-carbonate' soils having no appreciable B horizon. Rendzinas are often base-saturated (having a pH of 7 or more) with an upper horizon that is usually dark brown, sometimes with a whitish tinge on chalk. This grades directly into lighter coloured parent rock. Rendzinas are usually shallow, freely drained and carry a typical vegetation.

The nature and quantity of the organic litter in soil depends largely upon the type of vegetation cover and its rate of decomposition. In alkaline or neutral soils, decomposition tends to be fairly rapid as a result of the activities of bacteria and fungi which here attack both lignin and cellulose.

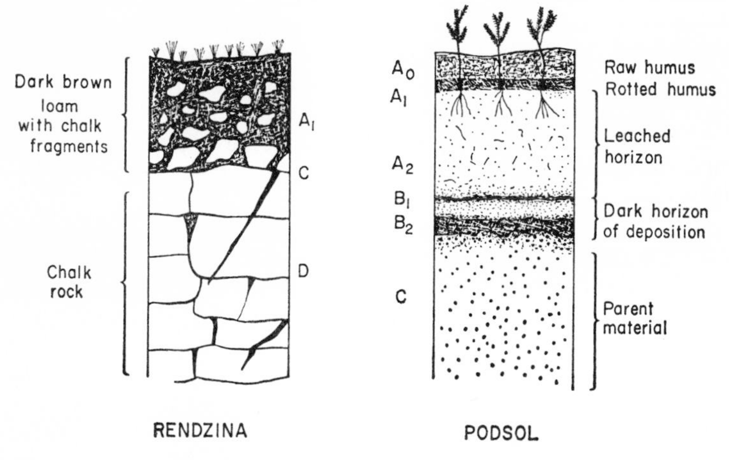

Fig. 1–2 Podsol and rendzina profiles.

At the same time, larger soil animals such as earthworms and insects contribute towards the mixing of humus with mineral particles. The humus of such soils is called *mull*, and is typical of brown earths and usually derived from broad-leaved trees. Intermediate between mull and the humus of greater acidity called *mor*, is *moder*, which has a richer and more varied fauna than mor, although plant remains are not broken down to the same extent as in mull. Moder tends to be considerably eaten into and mixed with faecal matter, yet is not matted together as raw humus and some mineral matter is also incorporated.

The pedocalic type of soil most similar to fully leached soils is called *chernozem* or *black earth*, and is found under 'steppe' and savannah grassland in central Europe and Asia, North America, Argentina, East and South Africa where, in spite of summer drought, the annual rainfall is over 25·5 cm (10 in.). The surface horizons appear dark due to the presence of organic matter which has become humified under conditions of high temperature and alkalinity. On the arid side of the chernozems a soil type known as *chestnut earth* develops. The humus horizons are less well developed than in chernozem soils, and the calcium carbonate deposits come near to the surface. Chestnut earths are covered with low grass steppe vegetation and occasionally some scrub.

In semi-arid regions where the rainfall varies between about 12·0 and 25·0 cm per year, the soil types usually formed are *brown* or *grey semi-desert soils*. The latter are sometimes called *sierozems*. Sierozems may have even less than one per cent of organic matter in the surface horizon and calcium carbonate deposited on the soil surface. They often support low desert-scrub vegetation.

Desert soils, produced almost entirely by physical weathering, contain no humus and are little more than fragmented rock. The winnowing effect of the wind, sorting out particles of different sizes, transporting and depositing them elsewhere, results in the formation of three main types of desert: *hammada* or rocky desert, *reg* and *serir* or stony desert and *erg* or sandy desert. Sometimes the finest soil particles are removed so far by wind that they are deposited in the *loess* of the steppe-lands that border the desert.

Low-lying regions close to the equator where the annual rainfall exceeds 200 cm (80 in.) are usually clothed in dense *rain-forest* whose determining conditions are a high, even temperature and abundant moisture. For the most luxurious development of rain-forest the precipitation must be distributed evenly throughout the year. In areas where the soil is periodically submerged, swamp jungle develops and this may in coastal regions take the form of mangrove swamps, for mangroves can flourish only where their roots are periodically submerged and uncovered by tidal action. In tropical and equatorial rain-forests (Plate 2(a)) there is little humus because high temperatures and humidity result in the rapid decay of organic matter. At the same time, lack of calcium owing to leaching possibly results in a poor invertebrate fauna, apart from the ever-abundant termites and ants which, by mixing humus with the soil, partially fulfil the functions of earthworms in temperate forests. In spite of the large quantities of vegetable matter consumed by termites, there appears to be little increase in soil fertility as a result. This is probably because the termites' digestion is very efficient and they eat not only their own excrement but also the corpses of dead termites. Thus little is left over to enrich the earth. Nevertheless, they move a great deal of soil and influence it considerably by their burrowings.

Soil profiles in part reflect features of surface relief as well as of rainfall and parental material. Shallow soils develop in hilly regions with accompanying excessive run-off and erosion. Flat land has little or no erosion so that a leached upper soil overlying a dense clay pan results.

Waterlogged horizons are usually greyish-green or greyish-blue indicating anaerobic conditions and the presence of organic matter. Such soils are known as *gley* and are often overlain by a horizon with rusty-brown mottling due to the presence of oxidized iron compounds. The layer affected is sometimes designated by the letter G or by (g) following the symbol for the appropriate horizon—e.g. B (g). Low-lying regions with poor drainage favour the accumulation of humus which may even form *peat*, containing more than 65 per cent by weight of organic material. In such anaerobic conditions, organic remains of plants become only partially decomposed.

Soil profiles can be studied from specially dug pits or, more simply, by the use of a *soil auger* or *soil sampler* (Fig. 1–3). The first is a steel instru-

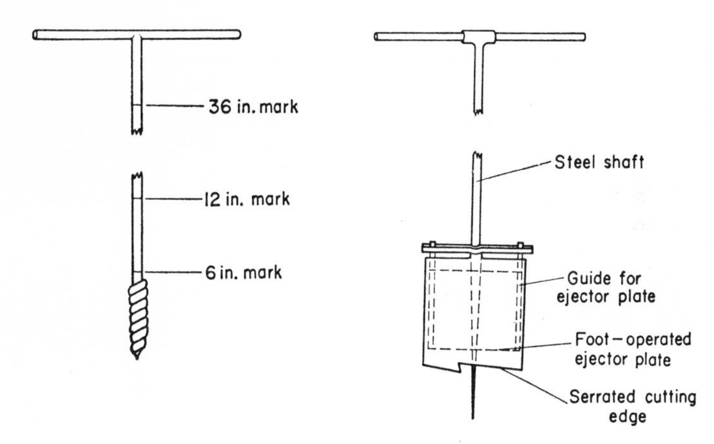

Fig. 1–3 Soil auger and soil sampler.

ment with a screw at the base. In use it is screwed 6 in. into the ground and pulled out. After the first 6 in. of soil have been examined and discarded, it is put into the same hole and screwed in to a depth of 12 in. The process is again repeated to a depth of 36 in. A soil sampler consists of a cylindrical corer which is forced into the ground. When filled with earth, it is removed and the sample examined directly or placed in a bag for subsequent study in the laboratory. The process is then repeated and the hole gradually deepened. Slopes can be measured by means of a *clinometer* (Fig. 1–4) consisting of a protractor whose base is aligned with the ground whilst a weighted string indicates the angle from the vertical. This clinometer illustrates the general principles involved, but would be far more practical if mounted on a wooden frame.

1.3 Plants

Land plants provide innumerable microhabitats. Many small animals use plants as shelter and protection from enemies, evaporation, wind, extremes of temperature and other unfavourable conditions. Many species of insects and spiders commonly rest inside flowers and seed pods. Leaf-rollers and other insect larvae fasten leaves together to make nests. Ants of the tropical genus *Oecophylla* co-operate in holding quite large leaves together whilst they are being bound in place with the silk produced by the larvae which are used as living shuttles (Plate 1).

In tropical forests (Plate 2(a)), spaces between the leaves of epiphytic orchids and bromeliads are inhabited by a vast array of insect larvae, planarians, earthworms, snails, woodlice, centipedes, millipedes, grasshoppers, earwigs, ants, scorpions, spiders, tree-frogs, lizards, snakes and

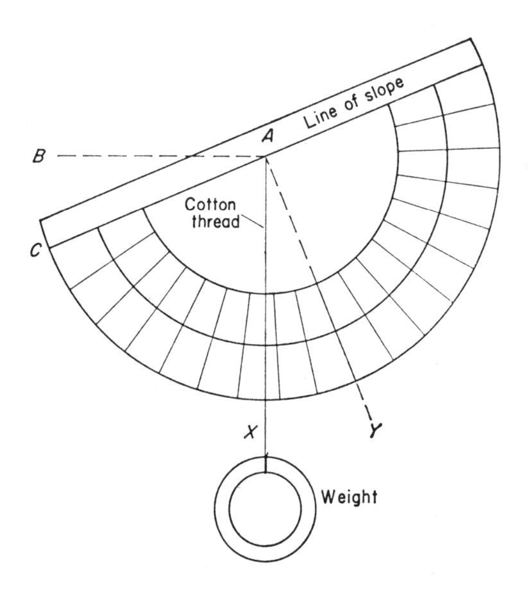

Fig. 1–4 Clinometer (angle of slope, $B\hat{A}C = X\hat{A}Y$) (after SANKEY).

so on. The bases of palm fronds similarly harbour small animals, many of which have evolved a flattened shape as an adaptation to their specialized environment. Leaf-miners are tiny insects whose larvae live and feed between the upper and lower epidermal layers of leaves. They also tend to be conspicuously flattened and have reduced legs. Some are restricted to a single plant species, others feed upon plants of unrelated genera. Even closer associations occur where plant-boring insects and mites stimulate the formation of galls in which they live.

One of the most interesting symbiotic relationships between plants and animals are those of the ant-plants. In tropical Africa and America, many *Acacias* have hollow stipular thorns which serve as dwellings for ants, and glandular secretions, which provide their food, are produced near the tips of the leaves. *Cercopias* are provided with many small chambers in

Plate 1 Leaf nest of the ants *Oecophylla* sp. in Zanzibar (see pp. 8, 25).

Plate 2 (**a**) A clearing in equatorial rain-forest, Uganda; (**b**) microclimatic measuring apparatus: *Left*. Cobalt thiocyanate paper comparator outfit. *Lower right*. Thermistor electric thermometer. *Upper centre*. Whirling hygrometer and micro-anemometer.

their stems and food is present on the petioles. The ants drive away leaf-cutting and phytophagous insects as well as browsing animals and, in return, are provided with food and shelter. One ant-plant in British Guiana was found to have no less than 50 kinds of animals associated with it, including 20 species of ants.

Some plant species have associated with them a much richer fauna than others, but the reasons for this are not really known. A high transpiration rate from the leaves may favour small animals and their predators and parasites, as may richness in calcium and other inorganic ions. The position in which a plant is growing may also markedly influence the animals that are associated with it. The fauna of nettle leaves, for example, varies according to whether the nettles are shaded or growing in sunlight.

Even the proximity of plants may create favourable microclimates for certain animals. For example, the mosquitoes *Anopheles billator* and *A. homonculus* of the Trinidad rain-forest concentrate at different heights and at different times of day. Air is moister nearer ground level, but the humidity gradient varies during the day, so that the mosquitoes move up or down and thus remain in a fairly constant micro-environment. *A. homonculus*, however, always remains nearer the ground than *A. billator* and is active during the moister times of the day.

1.4 Tree-trunks, logs and rocks

Tree-trunks, fallen logs and rocks provide innumerable microhabitats for small animals. Crevices and holes in the bark of trees and logs, and cracks in rocks often shelter a variety of creatures which thereby obtain shelter and protection, while many animals gather underneath rocks and logs. Both living and dead trees are attacked by wood-boring insects whose burrows form microhabitats in which their owners live.

Tree holes are usually small or moderately sized and often become filled with water in which breed mosquitoes and other aquatic insects. As fungus decay sets in, they tend to become enlarged and drier, and attract the fauna that normally occurs in rotting logs. Eventually the whole tree may die and decay, itself becoming a rotten log which continues to decompose until eventually it becomes incorporated in the leaf-litter stratum of the forest floor.

The fauna of rotting wood changes with the varying stages of decay so that a *succession* of animal communities develops, each one dependent upon a particular type of microhabitat. Faunal successions are both easy and interesting to study. For example, a newly-felled log can be examined at weekly intervals and changes in the fauna noted through the various stages of decay. After the larvae of wood-boring beetles have burrowed into the wood, carrying with them wood-rotting micro-organisms, the log becomes thickly populated by cryptozoic animals (p. 27) while woodlice, centipedes, spiders and even small vertebrates seek shelter under the peeling bark. Eventually numbers decrease as the wood decomposes to form humus.

Faunal successions can also be studied in marine habitats. A cubical cement block may be placed on the sea-shore, or parts of existing jetties cleaned and sterilized. The development of organisms on sides facing different points of the compass can then be studied. A protozoan culture made by enclosing rotting leaves in pond-water in a covered jar constitutes a microhabitat in which successional changes may be observed and measured.

1.5 Leaf-litter

Accumulations of fallen leaves and humus provide a habitat in which dwell large numbers of interesting animals. As a subject for ecological investigations, the fauna of leaf-litter has much to commend it. Leaf-litter varies greatly according to the vegetation that produces it which, in turn, is determined by the soil type. Gradually it decays and becomes incorporated in the humus of the soil so that there is again a succession of changes in the fauna and flora as also occurs in large-scale habitats.

A thick cover of litter tends to conserve moisture and equalize temperatures. Many litter animals are adapted to special or peculiar conditions: some mites will pass a whole generation inside fallen spruce needles after these have first been attacked by fungi. Casual investigation reveals little of the teeming life among fallen leaves. Many forms are not easy to perceive, most are agile and difficult to catch and nearly all avoid the light and quickly hide themselves when disturbed. Yet, for the ecologist using the methods described below, there are few more rewarding microhabitats.

1.6 Carcasses and dung

Faunal successions in the microhabitats formed by carcasses and dung have not been extensively studied, but, like logs, they are interesting because they are among the few known examples of primary zoological successions. Most zoological successions are secondary because they follow in the wake of vegetational changes. In temperate regions, the first animals to visit carrion are blowflies, *Lucilia* and *Calliphora* spp., which feed and lay their eggs. The maggots which hatch break down the tissues of the carcass and let in air, thus encouraging putrefaction. They are preyed on by histerid and staphylinid beetles and parasitized by braconids, *Alysia* spp., until they pupate, after which they are attacked by the wasp *Mormoniella* spp. Finally, as the carcass dries and becomes less smelly, it is devoured by larvae of the tortricid moth *Peronea* sp. and of Diptera such as *Piophila* sp., Phoridae and by dermestid beetles. Of course, this is only one of many successions that may take place depending on whether the corpse is in a damp place or dry, in sun or shade and so on. In desert regions, by the time kites, vultures, jackals and hyaenas have had their fill, even quite large corpses dry out and may remain like mummies for several months.

Cattle droppings also pass through various stages, associated with progressive loss of water, until eventually they lose their identity and become homogeneous with the humus of the soil. In temperate regions these stages or *microseres* may be quite prolonged and associated with a characteristic community of animals (BRADY, 1965) but, in the tropics, changes are much more rapid. Sometimes the droppings are removed by dung-beetles before other animals have had an opportunity to enjoy them. In deserts they usually become dry and hard so quickly that they do not attract insects, but remain in a desiccated state until they are fragmented and dispersed by the wind.

1.7 Walls and caves

Caves, holes in rocks and stone walls differ from the microhabitats previously described in that they do not show successions other than seasonal changes, nor do they provide quantities of vegetable matter as food. Apart from caves, they tend to be drier and less well protected so that their microfauna is far less rich, although snails and woodlice are often plentiful. When leaf-litter and soil accumulate a typical soil and litter fauna develops. Caves are especially important in desert regions for the shelter they provide (section 3.5).

On the sea-shore, caves provide an important microhabitat for marine and littoral organisms, for many animals found living between tide marks are limited to small caves and rock crevices. The marine animals such as molluscs and polychaete worms that depend upon dissolved oxygen and the semi-marine, air-breathing arachnids and insects are equally dependent upon a moist, sheltered microhabitat. The distribution of both marine and terrestrial components of the fauna is greatly influenced by tide levels and degree of exposure, however, and both show well-defined upper and lower limits.

1.8 Nests of social insects

The nests of termites, bees, wasps and ants provide relatively specialized habitats, inhabited by large numbers of small animals that obtain protection and shelter in the comparatively favourable microclimates therein. They have been studied in considerably more detail than have the microhabitats described above in sections 1.3 to 1.6.

The various types of organisms commonly found in close association with social insects have been classified in five ecological categories, viz. *synechthrans* or persecuted predators, *synoeketes* or tolerated scavengers *trophobionts* such as aphids which are tended by ants, usually outside their nests, for the sugary secretions they produce, *symphiles* or true guests of the colony which give exudates to their hosts who feed and guard them, and *parasites* both external or internal. Various types of social parasites and unwelcome guests are often lumped together under the name *inquilines*.

1.9 Summary

Ecology is essentially a practical subject and is best learned by study in the field. In this chapter I have outlined some of the more important microhabitats of small animals but it is obviously not possible to be exhaustive.

Anyone who goes into the countryside and uses his eyes will immediately spot many others. For example, the fauna of birds' nests, fox holes, rabbit warrens, lichen on walls, cellars and empty houses, grain stores, pitcher-plants, mossy banks and spiders' webs are all worth study. The sea-shore too, provides a number of exciting possibilities, some of which I have briefly mentioned. The faunas of sandy and muddy beaches can be compared and related to particle size, oxygen deficiency and other factors, whilst rocky shores provide innumerable isolated microenvironments varying in degree of exposure, solar radiation, salinity and so on. Even empty barnacle shells have their own characteristic fauna and provide, for example, a place of refuge for tiny dipterans inside the bubble of air, trapped when the tide rises.

Microclimatic Measurements 2

It has long been known that organisms occupying the same general habitat may actually be living under very different physical conditions. Only recently, however, has much attention been directed toward the detailed study of microhabitats and their climates. Investigation of microclimatic conditions in the past has often been hampered by lack of suitable apparatus, but recent innovations in the methods of measuring physical conditions in small enclosed spaces have largely obviated this difficulty. Even so, comparatively few attempts have yet been made to relate microclimates directly to the fauna inhabiting them, although there are a number of studies of microclimatic conditions in various types of plant cover. In this chapter I shall consider only simple methods that can be carried out easily in the field and do not require expensive or complicated apparatus.

2.1 Temperature

Two factors have to be taken into consideration in the design of microclimatic apparatus. Firstly, the sensory elements of the instruments to be used must be sufficiently small to be inserted into small spaces and, secondly, it should be possible to take readings at some distance from them. Thus, although a normal mercury thermometer can be used for some purposes, such as measuring temperatures below the surface of the soil or in soft, rotting wood, it has a number of drawbacks. Its large size is a disadvantage because there may be steep temperature gradients near surfaces so that a smaller sensitive element is essential. Then there is a tendency for heat to become conducted from the bulb along the stem so that the thermometer itself affects the microclimate that it is supposed to be measuring. Finally, the proximity of the sensing and reading parts of the instrument limit its use and it cannot be used to measure air temperatures when exposed to a source of radiant heat.

In contrast, thermocouples have the advantage of small size and are both cheap and easy to make. However, they have to be used in conjunction with a galvanometer which is an expensive and delicate instrument and, like the Thermos flask, liable to destruction in the field.* Moreover, the calibration of thermocouples is affected by the length of the leads because the resistance of the system is low and amplification is necessary if recording apparatus is to be used. A fairly simple battery-operated amplifier for continuous recording has been devised (DENTON, 1951), but all the objections I have mentioned can more easily be overcome by using a thermistor resistance thermometer instead (Plate 2(b)).

Thermistors (Fig. 2–1) make use of a semi-conductor mixture of oxides and metal, specially treated, and take the form of small glass beads less than

* see note on p. 19.

a cubic millimeter in volume on the end of a thin glass rod. They are used in a Wheatstone bridge circuit in conjunction with an ammeter, which may

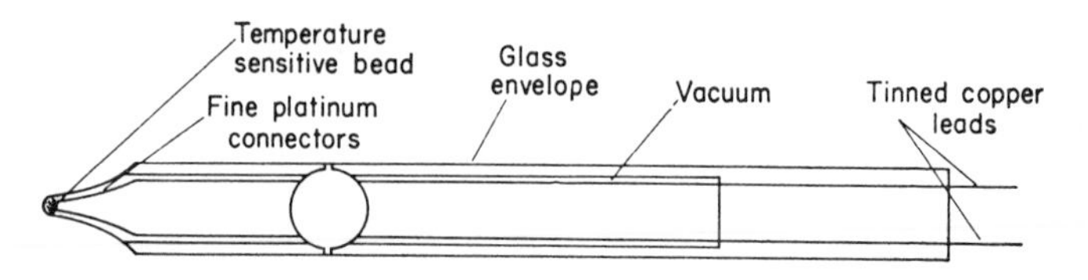

Fig. 2–1 A thermistor in section (after MACFADYEN).

be quite a cheap one, and a small dry battery (Fig. 2–2). The apparatus can be constructed very easily and is calibrated with a mercury thermometer in a water-bath. Very long leads are possible because the resistance of the element is high, 1,000–100,000 ohms according to type, and power demands are low so that an amplifier is not required. Consequently it is possible to construct a battery-operated apparatus that records the temperatures of 12 thermistors once an hour for a week (2,304 recordings) on little over a metre of film (KEMPSON and MACFADYEN, 1954). Many experiments, however, require bioclimatic measurements and faunal observations to be made at the same time: automatic recording apparatus is then of little use.

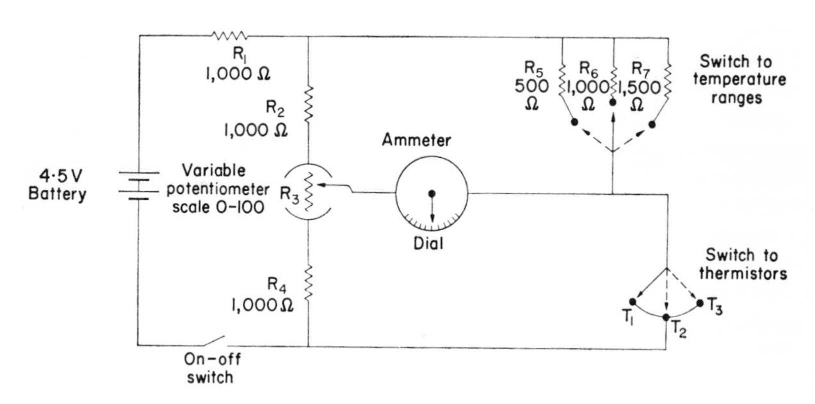

Fig. 2–2 Circuit diagram for themistor thermometer with three temperature ranges and three alternative thermistors (after CLOUDSLEY-THOMPSON).

2.2 Humidity

Methods of measuring humidity that depend upon the rate of evaporation of water are unsuitable for use in small spaces because the humidity to be measured would be influenced by the apparatus. For this reason hair- and paper-hygrometers have often been used for the measurement of relative humidity in small spaces. They require constant calibration, however, and are useless under extreme conditions which are often those of greatest interest in ecology.

A thermohygrograph which combines a bimetallic strip thermometer and a hair-hygrometer has been constructed from a watch, the hour hand of which is replaced by a light turn-table. On this are mounted two levers, one actuated by the bimetallic strip and the other by a humidity-sensitive hair. The tips of the levers bear fine points which project upwards and trace a circular line on the smoked watch glass. The glass is afterwards examined under a microscope; the distance from the centre of the watch is proportional to the factor to be measured. Accuracies of 0·1°C and 5 per cent relative humidity are possible within the range 30 to 90 per cent relative humidity (KROGH, 1940).

Simpler, and more reliable, are electric hygrometers, which consist of specially impregnated fabric carried on platinum electrodes supported by a plastic frame. This absorbs or gives off moisture and rapidly attains equilibrium with the surrounding air. The amount of moisture it contains governs its electrical resistance so it can be used in a Wheatstone bridge circuit just like a thermistor. The only drawbacks to this type of apparatus are the necessity for recalibrating it and the need to protect the element from dust and moisture (EDNEY, 1953).

Without doubt the most practical way of measuring humidity in very small spaces is to expose small pieces of paper soaked in a solution of cobalt thiocyanate. Cobalt thiocyanate paper is prepared commercially and readily obtainable. It changes colour from red to blue as the humidity decreases. Papers exposed in an environment in which humidity is to be determined are quickly mounted in liquid paraffin between a piece of colourless glass and a small white tile. They are later matched in a comparator with standards made by exposing papers to known humidities or with standards of coloured glass (Plate 2(b)). The papers may be very small in size and relative humidity can, it is claimed, be estimated to an accuracy of about 2 per cent (SOLOMON, 1957) but one of 5 per cent is easily attained.

Sets of standard colours can also be made with water-colour paints and by this means the whole comparator set up can be eliminated, the cobalt thiocyanate paper being compared directly with the water-colour standards. The only problem is to get known relative humidities from which to obtain cobalt thiocyanate paper standards in the first instance. Controlled humidities can be obtained by means of solutions of sulphuric acid, potassium hydroxide or sodium hydroxide, etc. Of these, potassium hydroxide or sulphuric acid solutions are perhaps the most convenient and the necessary data are given in Table 1. (Take care in diluting concentrated sulphuric acid or dissolving potassium hydroxide, as considerable heat is generated in the process.)

Table 1 Sulphuric Acid and Potassium Hydroxide Solutions for Control of Atmospheric Humidity at 20–25°C (after SOLOMON, 1951, *Bull. ent. Res.*, **42**, 543–554)

Relative humidity per cent	*Wt per cent* (g H_2SO_4 per 100 g solution)	*Density* (g/ml.)	*Wt per cent* (g KOH per 100 g solution)	*Density* (g/ml.)
100	0	1·00	0	1·00
95	11	1·07	7	1·06
90	18	1·12	12	1·11
85	23	1·16	16	1·15
80	27	1·19	19	1·18
75	30	1·22	22	1·21
70	33	1·24	25	1·24
65	36	1·27	27	1·26
60	38	1·29	30	1·29
55	41	1·31	32	1·31
50	43	1·33	34	1·33
45	45	1·35	36	1·35
40	48	1·37	38	1·38
35	50	1·40	40	1·40
30	52	1·42	42	1·43
25	55	1·45	45	1·45
20	58	1·47	47	1·48
15	61	1·50	50	1·51
10	64	1·55	—	—
5	70	1·60	—	—

NOTE: A simple method for controlling relative humidity is to make up a stock solution of equal volumes of concentrated sulphuric acid ('AR') and distilled water. Most of the heat produced by dilution of the acid is evolved during the initial stages so that, once a stock has been prepared, any subsequent dilutions can be made simply and quickly.

Relative humidity per cent	*Stock* ml.	*Water* ml.	*S.g. of mixture*
20	709	114	1·49
30	686	226	1·41
40	539	306	1·38
50	514	420	1·33
60	374	396	1·29
70	348	510	1·25
80	294	640	1·19
90	161	712	1·12

(After BUXTON, P. A. and MELLANBY, K., 1934, *Bull. ent. Res.* **25**, 171–175).

2.3 Soil moisture

In deserts, heaths, sand-dunes and other dry places, the soil humidity may be quite low. In temperate climates, however, the air between soil particles and in leaf-litter is usually saturated. Soil moisture can be measured by a number of methods of varying complexity. The simplest is to dry a known amount of soil at 100°C until no further loss in weight occurs and then to note the amount of water that has been lost in evaporation. More complicated and time consuming techniques involve the removal of samples to the laboratory and the measurement of the suction pressure of the water in them.

The soil tensiometer provides a measure of the soil moisture available to plants. This, of course, is not the same as the total soil moisture because it depends upon the mechanical composition and other factors of the soil, as well as on its osmotic pressure. Basically, a soil tensiometer consists of a porous porcelain pot containing water, to which is attached at the open end a mercury manometer or a vacuum gauge. As water passes in or out of the porous pot, the change is registered on the meter. This instrument is not sensitive in dry conditions.

Another method of measuring soil moisture depends upon measuring the resistance developed in a block of plaster of Paris, gypsum, nylon or fibre-glass supplied with two electrodes. The leads from the electrodes are plugged into an alternating current bridge when readings are required and continuous recordings obtained over long periods. The blocks are calibrated against a standard apparatus (BOYOUCOS and MICK, 1940).

2.4 Light

In general, the only satisfactory way of measuring light intensity is to use a photoelectric cell. Doubtless, both intensity and colour content of light are of significance to animals, but there are few studies on this problem and no special micromethods have been developed. A photometer has, however, been designed for measurements of an average light intensity in woods and other places where uneven distribution of light necessitates an integrating instrument with a sensitivity which is proportional to the intensity. It is based on a number of cells fitted with different filters (KOIE, 1954).

2.5 Wind speed

No special microclimatic methods have yet been evolved for studying wind speed in small spaces. Wind-vane and cup anemometers are scarcely applicable because of their insensitivity and large size. Fan type anemometers (Plate 2(b)) are more sensitive, but are still too large for microclimatic work and are insensitive to very low air-speeds; furthermore, they are affected by the direction of the wind. Perhaps the most suitable apparatus yet devised is a thermistor anemometer in which a heating element surrounds the small, naked bead. Over a range of speeds up to about 5 m.p.h., the extent to which the element is heated above the ambient air temperature shows an inverse linear relationship to the air speed (PENMAN and LONG, 1949).

2.6 Evaporation rate

Evaporation rate depends on three factors: temperature, wind speed and the saturation deficiency of the air. These factors are, in turn, themselves dependent upon a number of other factors. Consequently, although a number of 'atmometers' or 'evaporimeters' have been devised, they do not agree in their readings. The best-known consists of a porous sphere connected to a reservoir containing a measured quantity of water which is drained as evaporation takes place from the sphere. A micro version consists of a sintered glass funnel, 1 cm diameter, connected to a capillary tube with water and graduated. The rate at which water is withdrawn from the capillary indicates the evaporation rate. The capillary and sintered surface must be kept horizontal but, even so, results cannot be expressed in absolute units and can be used only for comparison of evaporation rates in different environments.

2.7 Hydrogen ion concentration (pH)

Hydrogen ion concentration can only be measured accurately by means of cumbersome and expensive electrical apparatus and it is much more convenient in the field to make use of less precise colorimetric methods. Soil pH seldom extends beyond the range 4·0–8·5. This can be estimated by means of a multiple indicator whose colour changes in steps of 0·5 pH. Special test tubes are available with two graduations: the lower indicates the amount of liquid to be tested (distilled or rain-water in which some of the soil, etc., has been shaken up) and the upper the volume of indicator to be added. The two are shaken together and the resulting colour matched against a standard chart of pH values. Even more rapid measurements can be made with the aid of pH indicator papers.

2.8 Summary

Although many of the techniques described in this chapter give absolute values, others can only be used for comparing one microclimate with another. This does not matter very much, because it is these comparisons in relation to the faunas of the microhabitats under investigation that are of interest.

For example, it is immediately apparent that woodlice and centipedes tend to wander in the open at night on walls and tree-trunks during warm, still weather. But, when there is a strong wind blowing, they stay down their holes. Even though a cup-type anemometer can only measure wind speed near to the surface of the wall, nevertheless it will give an indication of conditions on that surface between one night and another. Simple counts on different nights over a selected area will immediately indicate an inverse relationship between wind-speed and the number of animals abroad. It is more interesting to establish the existence of such a relationship than to worry about the exact wind speed and evaporation rate experienced by the woodlice on the surface of the wall.

Footnote to page 13.
With portable mini-potentiometers the reference junction is incorporated within the instrument and maintained at ambient temperature rather than at 0 °C. The temperature difference between the junctions, as determined from the potentiometer reading, is added to or subtracted from the reference temperature according to whether the unknown temperature is above or below that of the reference junction.

Microclimate and Faunal Distribution 3

The evaporating power of the air is the most important physical factor of the environment affecting the distribution of cryptozoic animals. This is because small creatures have a very large surface in proportion to their mass; consequently, the conservation of water is the prime physiological problem of their existence. Temperature and relative humidity are closely related, however, and it is often not possible to distinguish between them in the field.

Worms, leeches, slugs and terrestrial arthropods such as woodlice, centipedes, millipedes, springtails and other soil dwellers, avoid desiccation by remaining most, if not all of the time in a damp or humid environment. Others, such as many insects, spiders, scorpions and mites, possess a thin epicuticular layer of wax which is relatively impervious to water vapour and thereby reduces to a minimum water-loss by transpiration (Fig. 3–1). Unfortunately, such a layer is also impervious to oxygen and carbon dioxide. A respiratory mechanism has therefore had to be evolved which permits gaseous exchange to take place while restricting water-loss to a minimum. The spiracles of insects and the lung-books of spiders, scorpions and other arachnids are normally kept closed by means of special muscles, and only when carbon dioxide in the body begins to increase are they opened to facilitate respiration. It can easily be shown experimentally by weighing that the rate of water-loss by evaporation from an insect or arachnid is greatly increased when 5 per cent carbon dioxide is present in the atmosphere, as this results in the respiratory apertures remaining fully open. Before moulting takes place a new wax-layer is secreted beneath the old cuticle that is due to be cast off, so that the ecdysis is effected with a minimum loss of water.

In addition to such morphological and physiological adaptations to life on land, insects and arachnids have evolved special excretory products—uric acid and guanine respectively—which are extremely insoluble. Consequently nitrogenous waste matter can be eliminated from the body in a dry state and no water is lost in the process.

The majority of cryptozoic animals are restricted to moist conditions, although these must not be too wet as it is almost as dangerous for them to become waterlogged as to be desiccated. Therefore such creatures are most abundant in well-drained soils through which they can move upwards or downwards by means of reflex behaviour mechanisms and thus keep in a favourable environment.

It is probable that the evolutionary transition of many invertebrate types from aquatic to terrestrial life may have taken place via the soil where aerial respiration is not associated with desiccation. Hence, the series—annelids, woodlice and 'myriapods', insects and arachnids—may correspond with the series of environmental changes represented by life in water, soil and air. Possibly the Apterygota, as well as those pterygote insects and arachnids that lack an epicuticular wax-layer and are therefore confined to life in the soil, may represent an intermediate stage in the evolution of terrestrial life (GHILAROV, 1958).

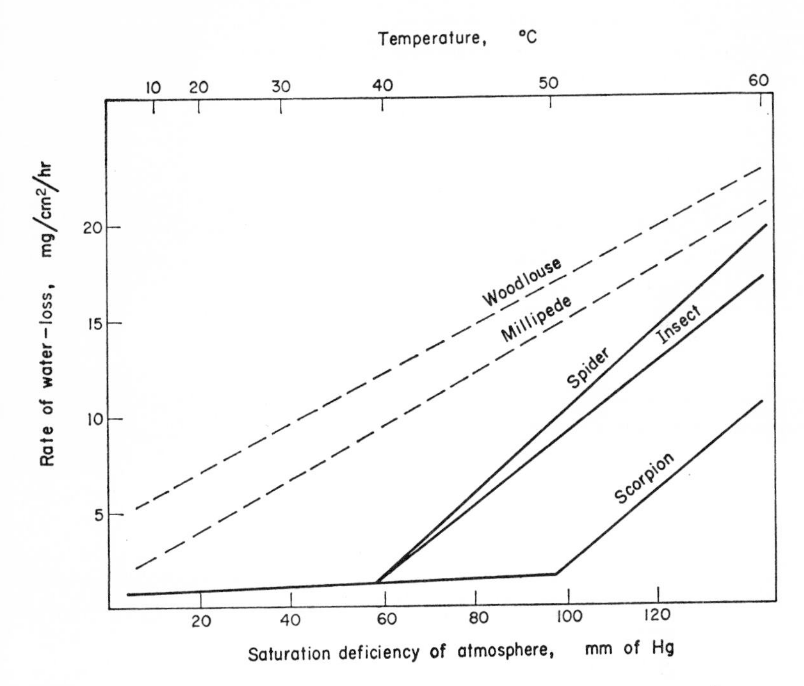

Fig. 3–1 Rate of water-loss in dry air at different temperatures and corresponding saturation deficiencies of a woodlouse, millipede, spider, scorpion and insect. In the woodlouse and millipede the rate of water-loss is proportional to the saturation deficit of the atmosphere; but in the insect, spider and scorpion it is negligible below a critical temperature at which their wax-layers becomes porous. Rate of water-loss is expressed in milligrams per square centimetre per hour (after CLOUDSLEY-THOMPSON).

3·1 Soils

The insulating effect of soil and leaf-litter has been demonstrated on a number of occasions: investigations on the soil temperature and its diurnal and annual variations have been carried out in many different places and a comprehensive literature exists. The soil receives practically all its energy from the sun so that temperature variation is closely related to solar altitude. Some of the incoming radiation is reflected from the surface, the remainder is used for heating and evaporation. Factors affecting this energy exchange include colour and slope of the soil, its heat capacity, moisture, drainage, wind speed and so on. A loose dry surface layer has a profound insulating effect which may be increased by the presence of humus and debris. Altitude may be important too, although it has been shown that burrowing to a depth of 6 in. into the soil may have a temperature effect equivalent to climbing 2,000 ft.

Sand temperatures decrease with depth and air temperatures with height during periods of strong or increasing radiant heating, but the temperature gradients are reversed when the rate of radiant cooling exceeds that of heating. Solar radiation is an extremely important factor governing sand temperatures. Surface temperatures over 60°C (140°F) have been noted in Finland and I have recorded a sand surface temperature of 82·5°C (182·5°F) in the Red Sea Hills in autumn when the air temperature fluctuated between only 40·5–43·5°C (105–110°F), and even more at Wadi Halfa in the Nubian Desert.

Diurnal temperature variation in the surface layers of the soil is considerable but decreases with depth and is almost zero below about 50 cm in most places. A diurnal range of 56·5°C has been recorded at a depth of 0·4 cm in Arizona. It so happens that the diurnal range is so nearly the same as the yearly range that the effects are not transmitted much below the surface and a relatively moderate and constant temperature is reached at a depth of 100–200 cm. For this reason it is important to know how deeply desert animals actually burrow to aestivate, hibernate or merely to pass the day.

Sand dunes represent an early stage in plant succession and a simple environment in that little is added to the soil stratum. It is, however, a severe environment; the organisms living in it are highly specialized and closely dependent on their microhabitats. The microclimates of sand dunes in temperate regions show many of the characteristics of desert sands. For example, the temperature at the surface among the Finnish dunes on a sunny day reaches a maximum of 47°C (116·5°F). At the same time the air 30 cm above the ground may be 29°C (84°F) and at a depth of 10 cm, only 17°C (62·6°F). Although surface temperatures may be high even in temperate regions, the relative humidity below the surface is not unduly low and the same is true of some deserts.

3.2 Plants

The presence of plants ameliorates climatic conditions to a marked degree. Not only the heights of plants but the quality and density of herbage affect temperature and humidity. Thus, whereas in young corn the hottest layer of air is at ground level and does not become remote from it until growth has proceeded for some time, in growing grass only a few cm in height the hottest air layer is already above ground level. The extent of the cooler, damper microclimate depends, too, upon the depth of penetration of the sun's rays and is determined by their angle, which varies diurnally and seasonally. Moreover, radiant heat penetration varies in grass which is partially flattened by previous high winds and thereby presents an upper surface of increased density. A further feature of the grass microclimate is the occurrence of temperature inversion during calm, clear nights when the air near the ground is warmer than that above. Such observations have been related to the behaviour of insect larvae such as Tenthredinidae. Shelter belts in steppe land have been shown to have an important effect in moderating high summer temperatures and the lethal effect of dry winds on hessian flies, *Phytophaga destructor*.

Not only may the temperature of one leaf vary quite markedly from that of another, depending on transpiration, relative humidity, and exposure to radiation and wind, but different parts of the same tree may vary in their capacities to respond to radiant heat. Staminate flowers of coniferous trees in Canada are commonly at temperatures 5–8°C above vegetative buds in sunlight, Since the buds, in turn, are above air temperature, this is an important difference for insects with habits like the spruce bud-worm, *Choristoneura fumiferana*. Parts of the same leaf even may vary in temperature so that in sunlight the centre of the leaf is several degrees warmer then the periphery, because heat exchange transmission between leaf and air is most rapid round the edges.

The temperature and relative humidity in summer among the roots of marram grass, *Ammophila arenaria*, among the coastal dunes of the Bay of Biscay have been found to be remarkably constant. These tufts of grass, particularly the older ones, harbour a number of smaller animals such as woodlice, spiders, mites and false-scorpions. The high relative humidity, 80–85 per cent, is due to the fact that at night dew runs down the leaf blades into the sand and humus at the roots.

3.3 Tree-trunks and logs

The temperatures beneath bark and in the galleries of wood-boring insects have been investigated by a number of authors and it has been found that different species tend to inhabit sunny or shaded regions. The development of bark beetles in logs also varies according to microclimatic conditions. In sectors most exposed to the sun, no eggs may be laid because daytime temperatures become too high and only in shaded sectors do the beetles develop normally. There may be a high death rate of larvae on the underside of logs when these become damp in wet weather. Recent studies of termite colonies in living trees in Australia have shown that the tree insulates the insects against fluctuating air temperatures in much the same way as a mound insulates colonies of mound-building termites. There was little diurnal variation in nests of *Coprotermes frenchi*, for example, and the temperature throughout the year varied only from 27–36°C. They were consistently higher than in the centres of uninfested trees (GREAVES, 1964). Microclimatic conditions underneath logs and rocks are often favourable for cryptozoic animals. Even in a wood in my garden in England, the range of temperature beneath a 3 in. thick stone was halved compared with that of the air, while on the upper surface the range was three times as great.

3·4 Leaf-litter, moss and debris

Convergent adaptations to repeated drying-up of the habitat appear frequently in the animal communities of moss and debris. Moss beds, in particular, may be regarded as small periodic bodies of water; the change from damp to dry conditions often occurs. Many of the inhabitants escape periods of drought by moving down into the soil below whereas others, such as tardigrades and rotifers, survive drought anabiotically, when metabolism ceases and they appear quite dead, or in *diapause* (section 4.5). It is clear that moss does not form a biotope with a stable microclimate and severe selection by unfavourable environmental factors prevents competition of its inhabitants with other local animals.

Leaf-litter and debris tend to produce much less extreme microclimates and support a rich and varied cryptozoic fauna. Even the thin drift on the seashore provides a microclimate within the range of tolerance of many animals and well below the high extremes of unprotected sand surfaces.

3·5 Caves

Caves form the natural habitat of many animals, and their microclimates are therefore of considerable biological interest. Not only are certain species confined to a cavernicolous environment, but others are only able to survive the rigours of the day in these cool, damp places. Special attention has been paid to caves in desert regions because these provide a wide range of microclimatic conditions, only a few metres from one another, from among which an animal can choose with little expenditure of energy.

For example, at a depth of 20 m inside a cave in dry rock near Cairo the annual temperature was found to be almost constant and the relative humidity was higher in summer than in winter, the reverse of what happens outside. In a cave at Jenin, Israel, the outer regions where daylight penetrates are hot and dry during the summer but at 7 m from the entrance, wet and dry bulb temperatures begin to approach each other and at 20 m a relative humidity of 100 per cent obtains. In temperate regions cave air tends always to be quite saturated with moisture.

3·6 Nests of social insects

Many of the terrestrial invertebrates that live in holes and burrows have evolved barricades or doors of various kinds which preserve the microclimatic conditions within, as well as serving as a defence against enemies. An extreme form of such niche adaptation is seen in the hole-closing devices of animals which employ parts of their bodies for the purpose. This phenomenon is termed *phragmosis* and is exhibited, in particular, by certain ants and termites in which the heads of some soldiers are modified to fit the openings in the woody plants inhabited by a colony. An example of phragmosis is also afforded by the larvae of tiger-beetles (Cicindelidae) which live in vertical burrows in sandy soil. Here they rest with the head level with the surface and closing the entrance.

It is well known that honey-bees are able to regulate the temperature of their hive. When the colony gets too hot, the bees cool it by bringing water, which is deposited on the upper parts of the combs. Then they fan vigorously with their wings and drive the air, cooled by evaporation, through the brood comb. On the other hand, in winter, the workers form a dense cluster which reduces heat loss and they can warm the hive if necessary by running about. This increases the rate of metabolism so that their bodies are warmed. To a lesser extent, bumble-bees are also able to control microclimatic conditions within their nests.

Microclimates within the temporary or bivouac nests of army ants, *Eciton* spp., have also been the subject of investigation and it has been found that an appreciable stabilization of microclimatic conditions occurs. These nests are formed wholly by the ants' own bodies clustered together, and generally hang from an overhead support, such as a fallen branch or log, to the ground; they vary up to about 25 cm in height to 50 cm in diameter. Control of intrabivouac temperature through the conservation or release of excessive heat, as well as buffering of the brood against external temperature fluctuations, is affected to a marked extent through the activities of the workers in the bivouac interior and its walls. In the dry season on Barro Colorado island, Panama, when the environmental temperature and relative humidity show their greatest fluctuation, the nomadic phase occurs in the hours after dusk (SCHNEIRLA, BROWN and BROWN, 1954).

Microclimates within the leaf nests of *Oecophylla* spp. (Plate 1) have not been investigated but it is known that microclimatic control is much greater in non-nomadic ants with permanent nests. Tropical termites, too, have to protect themselves from excessive insolation. The nests of some species are extended in a north–south direction so that a comparatively small area is exposed to the noonday sun. The earthen galleries which these insects build along their paths are also a protective measure against heat and excessive evaporation.

3.7 Summary

Just as plants and their habitats are closely related, so are animals and the microhabitats they inhabit. Although it is possible for animals to survive for a while in unfavourable circumstances, they differ from plants in their ability to move and seek out more favourable resting sites. For this reason, a study of microclimates and their influence on diurnal and seasonal rhythms of activity is of paramount importance in the understanding of animal ecology. The mere accumulation of data, however, will never solve problems. There is a great need for the distribution of animals to be studied in relation to their microhabitats and not merely the microclimates to be studied without reference to the organisms living within them (CLOUDSLEY-THOMPSON, 1962).

Cryptozoic Animals 4

As their name indicates, *Cryptozoic* animals (Fig. 4–1) lead hidden lives. Nearly all the animals that inhabit the microhabitats described in the previous chapters are cryptozoic and they are mostly rather small. However, apart from Protozoa, which are not included in this Study, most of them can be seen with the naked eye even if a lens or microscope is needed to make out their anatomical details.

4.1 Collecting

The quickest way of finding out what species are present in a particular microhabitat such as leaf-litter is to rake some of it together, put it in a sack for transport and take it back to the laboratory. Small quantities are then placed in a coarse-net sieve or on a piece of fine-meshed wire netting and shaken over newspaper or a white cloth. A wooden table painted white is even better. As the animals fall from the sieve, they are easily seen if the surface is brightly illuminated. They usually include worms, slugs, snails, woodlice, centipedes and millipedes, spring-tails, bristle-tails, small flies, beetles, mites, spiders, harvest-spiders and occasionally false-scorpions. The majority are very susceptible to desiccation and should either be killed immediately in 70 per cent alcohol or else placed in glass containers lined with damp filter paper to keep them alive.

An *aspirator* or *pooter* (Fig. 4–2) is essential for picking up small invertebrates in large numbers. It consists of a short tube of Perspex or hard glass, corked at either end. Each cork is pierced by a short length of glass tubing. One is used for sucking up insects, etc., the other is attached to a length of rubber tubing ending in a mouth piece. A piece of gauze tied across its inner end prevents the contents of the aspirator from being sucked right through into the mouth of the collector. A strong sheath knife, chisel or screwdriver is most useful for levering up the bark of trees and fallen log so that the animals beneath can be collected. Those that are too large to be sucked into a pooter can be picked up with forceps. A camel-hair brush provides a convenient method of picking up very small creatures and popping them into alcohol. A tuning fork is sometimes useful for attracting spiders out of holes in stones and a trowel for digging into earth or dung.

A selection of small tins and unbreakable glass or plastic tubes for the reception of cryptozoic animals from different habitats should always be carried in the field. Some should contain 70 per cent alcohol but it is best to have others clean and dry if you want to keep certain animals alive for further study. The difficulty about field work is that if you take with you all

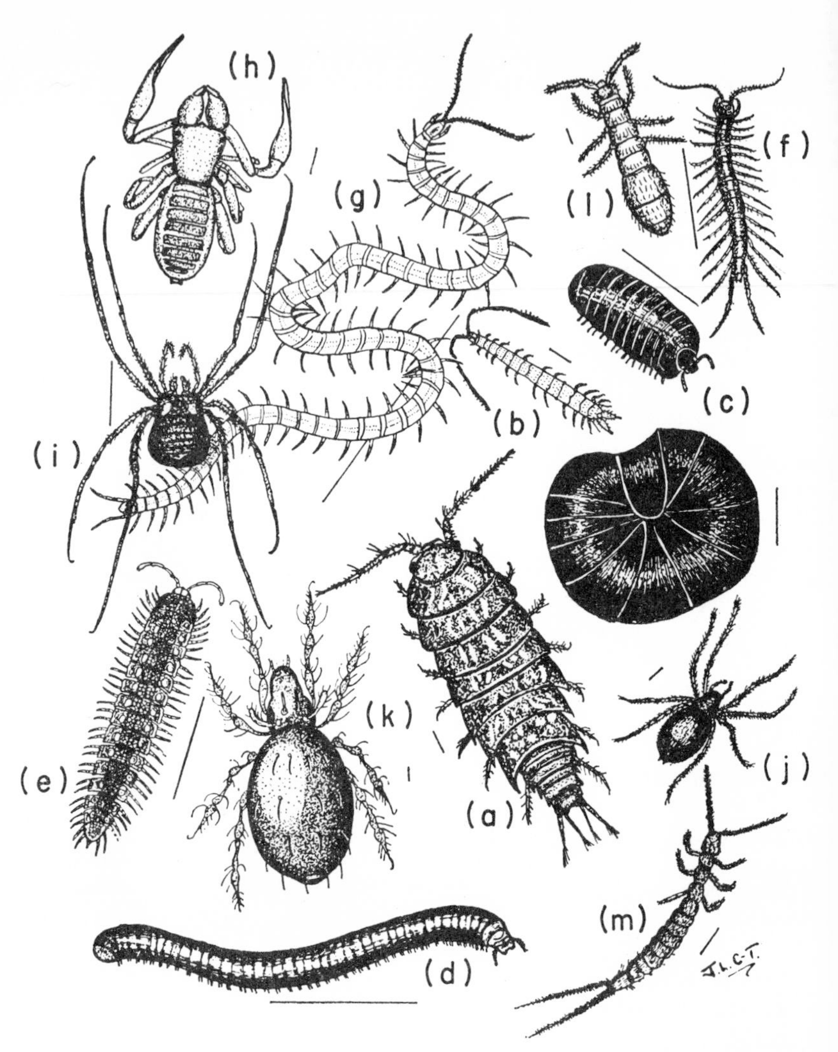

Fig. 4–1 Some typical cryptozoic animals: (**a**) woodlouse (Isopoda: Trichoniscidae); (**b**) symphylid (Symphyla: Scutigerellidae); (**c**) pill-millipede (Diplopoda: Glomeridae); (**d**) false-wireworm (Diplopoda: Iulidae); (**e**) flat-backed millipede (Diplopoda: Polydesmidae); (**f**) lithobiid centipede (Chilopoda: Lithobiidae); (**g**) geophilid centipede (Chilopoda: Geophilidae); (**h**) false-scorpion (Pseudoscorpiones: Chthoniidae); (**i**) harvest-spider (Opiliones: Nemastomatidae); (**j**) spider (Araneae: Linyphiidae); (**k**) beetle-mite (Acari: Oribatei); (**l**) spring-tail (Collembola: Poduridae); (**m**) bristle-tail (Diplura: Campodeidae). (Drawings not to scale: the straight lines indicate natural sizes) (Reproduced from *Country-side*, **17**, 460, the Journal of the British Naturalists' Association)

the apparatus you might need, you will be so cluttered up that it is difficult to carry it all even with the help of a haversack. It is therefore to be recommended that each person should concentrate upon only one or two kinds of animal or habitat at any one time. The minimum requirement is often only a knife or screwdriver. A moistened finger tip is an excellent substitute for an aspirator or forceps. In order to catch a tiny animal you just lick your finger and touch it, when it becomes stuck by the saliva. For quantitative methods of collecting soil animals, see *Life in the Soil*, a study in this series by JACKSON and RAW (1966).

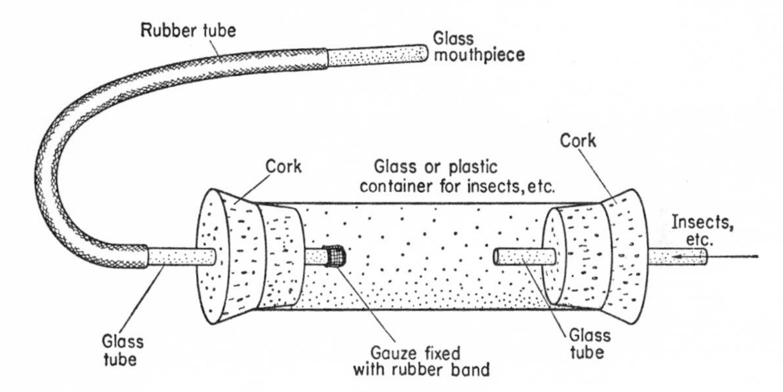

Fig. 4–2 Aspirator or pooter.

4.2 Identification

Although it is suggested in the Introduction that the beginner can, to some extent, evade the problem of identification by concentrating on autecological projects, sooner or later he will have to come to grips with it. Even if he does not intend to identify his catch in detail he will, at least, have to sort it into phyla, classes, orders and families and, in the case of common forms, into genera and species. In *Land Invertebrates* (1961), John Sankey and I have tried to provide a guide to the identification of common British worms, molluscs and arthropods other than insects, and we give references to other faunistic works. Lack of works of reference will make the precise identification of cryptozoic animals almost impossible to students in most tropical countries.

The beginner should familiarize himself with one group of animals before turning to others, as he will need to learn something of its morphology and nomenclature in order to understand the terminology employed in the keys for identification.

On returning from an afternoon's collecting among fallen leaves, the student will probably have a tube or two containing an assortment of small creatures in some muddy 70 per cent alcohol. Unless he intends to deal with them at once, the tubes should, of course, be labelled with locality and date before being stored. Let us assume, however, that he intends to work on his catch immediately. First of all, he will tip his tubes into a saucer or Petri dish and sort the animals into groups, each of which is placed in a separate tube with a label—the memory should never be trusted in such matters. You have only to lay aside an interesting collection of animals and, when they are rediscovered after a few weeks or months, all the necessary data will have been forgotten.

Some animals are comparatively easy to identify, others are extremely difficult.* The harvest-spiders are an example of the former and should present no problem. On the other hand, spiders vary enormously. Many of the larger species can be named with a fair degree of certainty after a few weeks' determined effort, but it will be years before the smaller Linyphiidae can be tackled with the slightest confidence. For, even where one has the necessary information to identify a specimen, it may be far from easy to see the diagnostic characteristics.

Immature spiders are a great puzzle to anyone beginning the study of the order, because identification usually depends on the shape of the male and female reproductive organs. A young spider may be recognized by its transparent appearance, sometimes of the whole body, but always of the male *palpal organs* which may appear rounded and swollen, but without the chitinous spines, apophyses and lobes characteristic of the adult state. The external genital organs or *epigyne* of the young female are semi-opaque and again have no chitinous outline distinctly marked. The best plan is to leave all immature spiders well alone.

Having now sorted the collection, let us select a spider for identification. Perhaps it is a large greyish one that was hiding under a stone among the tree roots on a bank. We take up a suitable reference such as LOCKET, G. H. and MILLIDGE, A. F. (1951, 1953) *British Spiders* (Vols. I and II) *Ray Soc.*, London, and turn to the 'Key to the Families'. 'Chelicerae very massive, articulated ventrally'—no, our specimen is clearly not the purse-web spider, for its chelicerae articulate obliquely and the spinnerets do not have three segments.

In the next couplet we find: 'Cribellum present with a calamistrum on metatarsus IV', *or* 'Cribellum and calamistrum absent'. Having refreshed our mind as to the meaning of these terms by glancing at the illustrations in the chapter on external anatomy, we hold up a rear leg with the forceps and look at it very carefully. No trace of a calamistrum can be seen, nor is there any cribellar plate in front of the spinnerets, so we can turn to the third couplet.

As the spider lies in the watch-glass we can see four eyes on the front part of the body. But the book says: 'Spiders with two eyes . . . Spiders with eight eyes' or, 'Spiders with six eyes'. How many has this one got? The small back ones are not easy to distinguish, so we twist the animal from

* The following remarks apply to the British fauna which is comparatively well known.

one side to another in the beam of light from the microscope lamp, until we can be certain. Definitely eight eyes are present. . . . And so the keys are followed, character by character, until the quarry is run to earth. Finally the epigyne or pedipalp of the spider is compared with the drawing in the book. If it agrees, this confirms that the spider has been correctly identified.

There can be few people who do not sometimes 'cheat' however, by turning to the illustrations first and then working the key backwards; or by comparing the spider with specimens already named. As it may take a long time to run it down by means of the key, and there may be several others awaiting attention, this procedure can be justified. Gradually, as one's collection of named specimens grows, identification becomes easier. But, as was said in the first place, it takes time and hard work even though it is fun. To begin with, one can never feel certain of anything: confidence comes gradually. And so it helps if one remembers that simple questions are by no means easily answered. For example, has a particular spider two or three claws on each leg? The tarsi are usually so hairy that the claws cannot be seen at all clearly. Actually, if three claws are not fairly readily visible, then it may be decided safely that only two are present. (If I had only known this when I first began to study spiders, how much trouble I should have been saved!)

4.3 Orientation

Most cryptozoic animals such as woodlice are in constant danger of desiccation and must remain in a damp moist environment. When taken into the open they are stimulated by light and drought so that they run actively until, by chance, they reach some dark, damp spot where they come to rest. They can run directly away from light, but one searches in vain to find any directed orientation towards moist air or damp surfaces. The animals run aimlessly, turning first one way then the other. But, as the air becomes damper, not only does their speed decrease, but they tend to turn less and less frequently until they finally come to rest.

Although this general hypothesis is now quite well established, there are many details still awaiting elucidation. If our knowledge of the biology of the inhabitants of leaf-litter and other microhabitats is to increase, advances must be made along three main lines; observation of behaviour coupled with precise measurement of microclimates in the field; laboratory analyses of the orientation mechanisms displayed; and an investigation of the sensory mechanisms involved. It is with the first of these in particular that we are now concerned. For further information, see *The Study of Behaviour* in this series by CARTHY (1966).

In discussing mechanisms of orientation, the modified terminology introduced by FRAENKEL and GUNN (1961) is usually adopted. This is as follows:

1. *Kineses.* These are the effects produced by stimuli on the rate of random movements of an animal. They are undirected, there being no

orientation of the axis of the body in relation to the stimulus. Kineses may be subdivided into:

(*a*) *Orthokineses.* Simple effects on the rate of locomotion depending on the intensity of stimulation. For example, woodlice move more slowly in damp than dry air and come to rest when the atmosphere is saturated.

(*b*) *Klinokineses.* The frequency of turning, or the rate of change of direction of movement, is dependent on the intensity of stimulation. Thus, when woodlice run into dry air, they start to turn around and run in circles. This often brings them back into damp surroundings where they cease to turn so frequently. On the other hand, if they do not immediately get back into a damp atmosphere, the rate of turning again decreases as they become adapted to less favourable circumstances. If the air now gets drier still, they again respond klinokinetically by constantly changing their direction while, if it gets damper, they start to run in a straight line.

It is by such purely mechanical reactions that woodlice aggregate under bark and rocks. Experiments have shown that the animals are not to any degree attracted towards one another or to favourable conditions. They merely gather in the same place as a result of behaviour reactions which drive them away from light and from dry places—a singularly negative but by no means inefficient method of orientation!

2. *Taxes* (*topotaxes*). There are movements which result from a discrimination of the direction of stimulation. They depend upon the presence of paired sense organs, except in the case of klinotaxis, and are thus more complicated than kineses.

(*a*) *Klinotaxes.* The direction of movement is dependent on the comparison of intensities of stimulation on each side by regular lateral deviations of a part or whole of the body; that is, by the comparison of intensities of stimulation which are successive in time. For example, if you were blindfolded so that you could still just discern light from darkness, by moving your head from one side to the other you could still find your way towards or away from the light.

(*b*) *Tropotaxes.* Movements whose direction is dependent on the comparison of intensities of stimulation on bilateral sense organs. Attainment of orientation is direct and no deviations are required.

(*c*) *Telotaxes.* Orientation to one stimulus as if it were a goal, disregarding all other sources of stimulation. Attainment of orientation is direct, without deviation, as when a dragonfly selects one insect from a swarm and catches it.

It must be remembered when attempting to analyse the behaviour of an animal that the mechanism involved may change from moment to moment and at the same instant more than one mechanism may be involved.

The object of experimental work in the laboratory should be to obtain fairly quickly certain information that might otherwise require years of field observations before it could be established. For example, millipedes are attracted to dilute solutions of sugar or, to be accurate, having by chance once come in contact with a sugar solution, they are afterwards repelled by its absence. This can easily be proved in the laboratory because they will aggregate on a filter paper soaked in dilute sucrose in preference to one soaked in distilled water. In the field, however, they seldom do this. Apparently they are repelled by the ants which are also attracted in vast numbers to the sugary filter paper.

I do not intend here to give precise details of how to test an animal to every kind of stimulus. Not only would this be extremely tedious, but each species requires its own techniques. The student must design an experimental method, based on his field observations, to suit the peculiarities of the species he is investigating. At the same time, as far as possible, he should investigate the kind of stimulus that the animal might be expected to encounter in its normal life. Thus, to show that a scorpion is repelled by the smell of oil of cloves may have little ecological significance, although this might prove to be a convenient stimulus to use for investigating the sense organs concerned in the reception of odours.

The chief stimuli in respect to which animals orientate themselves are: light, temperature, air humidity, moisture, gravity, contact, sound and other mechanical stimuli, airborne odours and contact chemical stimuli or tastes. Experimental technique consists in observing the behaviour of an animal in certain controlled conditions. For example, reactions to light can be tested by placing an animal such as a beetle in a darkened room and observing its reactions to a horizontal beam of light. Having seen whether it is photopositive or photonegative, the type of reaction involved can be determined by covering one eye with black paint or dark nail varnish and illuminating from above. If the animal then tends to run in circles, performing *circus* movements (Fig. 4–3), we know that the reaction is a tropotaxis. On the other hand, if it moves steadily towards or away from the light, turning its head first to one side then to the other, like a maggot, we know that the reaction is a klinokinesis. The intensity of response can be

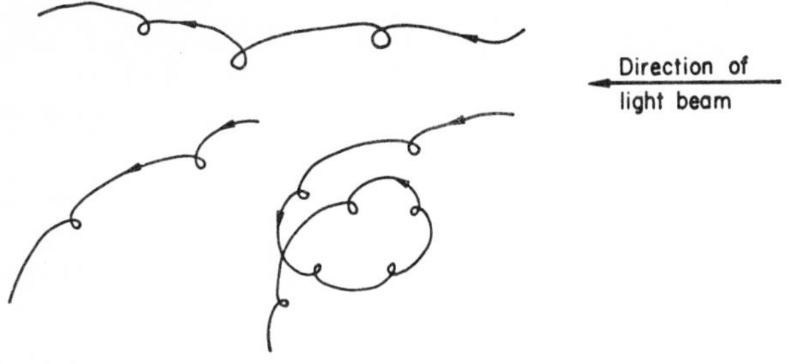

Fig. 4–3 Tracks of a photonegative insect, blinded on the left side, in a horizontal beam of light, to show circus reaction.

ascertained by counting the number of animals on the light and dark side of a choice-chamber (p. 36), one half of which is illuminated from above.

In experiments of this kind one has to use a little common sense. Some students once complained that the photonegative beetles provided did not respond to light or dark. They had left the insects in a glass beaker on the window sill for some time before testing their reactions to a dim light in a dark room. Of course the unfortunate creatures had been completely dazzled! Their normal reactions only returned after half an hour in the darkness.

Reactions to gravity may be established by seeing if an animal turns round when the vertical stick on which it is crawling is suddenly inverted. The experiment must be carried out in darkness or red light, of course, to eliminate optical orientation. The gravity responses of some cryptozoic animals can be determined by placing them in a vertical cylinder containing soil or leaf-litter and seeing if, after a period of time, they have crawled upwards or downwards. The method is also suitable for use in the case of stored product pests such as flour beetles and mealworms in bran. The technique can be improved by dividing the cylinder into sections held together by an outer sleeve, so that the animals in each section can be counted separately.

Temperature preferences can be investigated by means of a temperature gradient apparatus (Fig. 4–4). This consists of a long trough in which is placed a number of animals. The trough is heated at one end and cooled at the other and its inhabitants aggregate at their preferred temperature. This method, however, is open to two serious objections. First, any linear apparatus has end points where animals may congregate in response to contact stimulation. Secondly, some animals, like ticks, wander so far towards the cool end that they become inactivated there by cold. This gives the false impression that they have selected a temperature near to

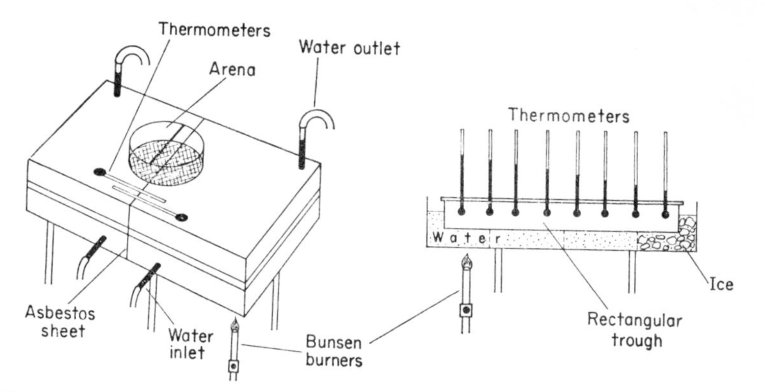

Fig. 4–4 Temperature choice-chamber and temperature gradient apparatus (in section).

freezing. For these reasons, choice-chambers are preferable to gradient apparatus.

The temperature choice-chamber that I use consists of two rectangular metal containers measuring 20 × 10 × 10 cm separated by an asbestos sheet 2 mm thick and held together by an outer metal strip. Warm or cold water can be circulated through the containers, or they can be heated from below. The temperature on their upper surfaces is measured by means of thermometers stuck down by plasticine. On the top of the metal containers is an arena composed of an inverted Petri dish. A glass rod fastened with paraffin wax across the top of the arena forms a partial barrier between the two sides, yet allows sufficient space for the animal to crawl unhindered underneath. The floor is of fine voile stretched over zinc gauze (Fig. 4–4).

One drawback of this type of apparatus is that temperature is not the only variable factor. Whenever there is a temperature gradient, there must also be a gradient of relative humidity and saturation deficiency. Most soil and litter animals, however, prefer a high humidity and this can be obtained by placing moist filter paper underneath the zinc gauze. The air will now be completely saturated on each side of the arena and the humidity gradient disappears.

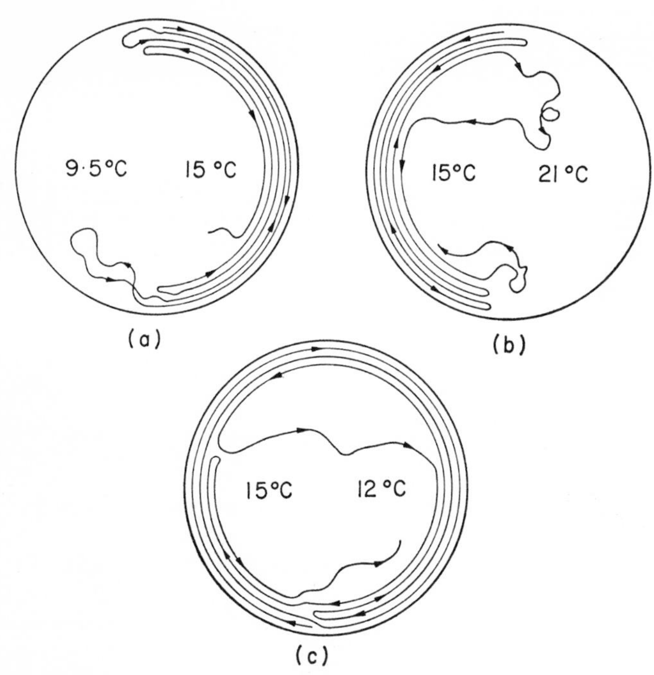

Fig. 4–5 Tracks of a small animal, e.g. a millipede, in a choice-chamber, showing klinokinetic reaction at boundary. (**a**) Avoidance of 9·5 °C; (**b**) avoidance of 21 °C; (**c**) no preference between 12 °C and 15 °C.

Choice-chamber apparatus can be used either with single animals or with groups. When a single animal is used, its path during periods of 5 or 10 minutes must be copied on a piece of paper (Fig. 4–5). Inspection of such tracks not only indicates preferred temperatures, but also reveals klino-kinetic reactions at the boundary between the two temperatures. Most animals run round the edge of the arena, responding by *thigmotaxis* to the stimulus of contact with it.

The relative intensities of response to various stimuli can be ascertained by inserting a group of 5 or 10 animals in the choice-chamber and counting the number that come to rest on either side of the arena after periods of 10 minutes. Between readings, the animals are thoroughly stirred with a glass rod and the apparatus twisted round so that the effects of external factors, such as light, are cancelled out.

A choice-chamber for investigating responses to relative humidity is illustrated in Fig. 4–6. In this case the gauze rests upon another divided

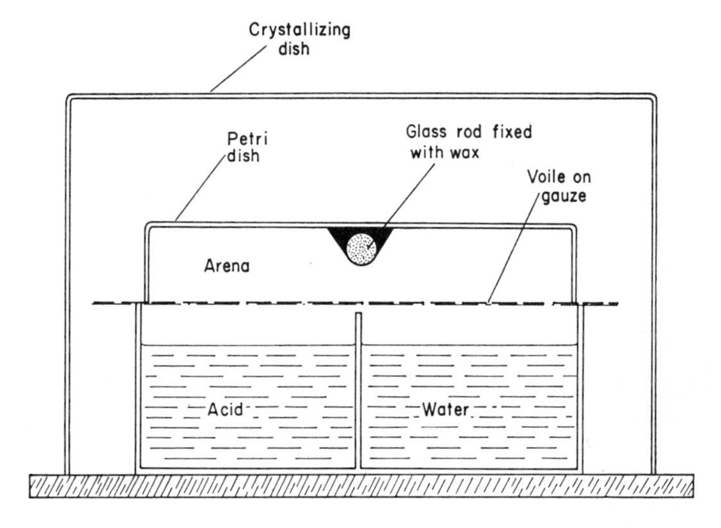

Fig. 4–6 Humidity choice-chamber apparatus (in section).

Petri dish, each half of which provides a different relative humidity in the half arena immediately above it. Controlled humidities can be obtained with the sulphuric acid or potassium hydroxide dilutions listed in Table 1 (p. 16) and the relative humidities within the arena checked with cobalt thiocyanate papers (section 2.2). The whole apparatus is covered with an inverted crystallizing dish to exclude draughts and sudden temperature changes.

Reactions to moisture can be tested by offering a choice of wet or dry filter paper on the floor of the arena, reactions to contact by offering different types of substrate and to taste by offering a choice of filter papers soaked in various solutions. The choice-chamber technique is capable of endless modifications. When the reactions of an animal to various stimuli have been evaluated, the influence of one factor upon another, such as the effect of light on the humidity response in woodlice (section 6.1) should be investigated.

4·4 Diurnal rhythms

The majority of cryptozoic animals are active at night and spend the day time in their sheltered microhabitats. Most of them lose water rapidly by *transpiration* and so cannot withstand the heat and drought of day in the open. They can only afford to come out at night when the temperature falls and the relative humidity of the air increases. At the same time, this is when they get rid of surplus water absorbed during the day (section 6.2).

The nocturnal habit confers several other advantages, even to animals that resist water-loss from transpiration. Enemies are more easily avoided, food more easily obtained and competition reduced. As a result of this, vulnerable and primitive forms especially tend to be nocturnal. Thus, among insects, silver-fish, bristle-tails, cockroaches, stick-insects and crickets are typically nocturnal. More highly evolved insects are quicker-moving, diurnal and live in hot, light environments. The same is true of spiders and other arachnids. Scorpions and Solifugae, for example, are primitive and markedly nocturnal, yet they are extremely resistant to high temperature, drought and other climatic rigours. So it is probable that their nocturnalism is correlated with ecological factors, such as the avoidance of vertebrate predators, rather than with physiological requirements.

It has long been realized that most animals are active only at particular times of the day or night but until recently it was thought that diurnal rhythms were *exogenous*, that is, simple responses to environmental changes. Laboratory experiments, however, have shown that most, if not all rhythms persist under constant conditions and are therefore *endogenous*. It is now generally recognized that activity rhythms represent self-sustained oscillations that are free-running under constant conditions and have their own inherent frequency which approximates to 24 hours—hence the term *circadian*, derived from the Latin *circa*, 'about' and *diem*, 'a day'. Under natural conditions, circadian rhythms of activity are synchronized with, or entrained to, the period of the earth's rotation by means of periodic factors of the environment. Of these factors, light is undoubtedly the most important. Experiments with many species of animal have shown that, with increasing intensity of illumination, the circadian period is shortened in day-active animals and lengthened in nocturnal forms. In this way circadian rhythms can be shifted to compensate for seasonal changes in day length.

In order to determine experimentally the time of activity of an animal, apparatus known as an *aktograph* (Fig. 4–7) is often employed. This consists essentially of an arena or box pivoted on a knife edge or on two needle points above its median transverse line. Any movement of an animal along the longitudinal axis tips the box and is recorded by a stylus writing on a smoked drum. The size and construction of the box can be adjusted to suit the size of the animal. For very small insects and other arthropods, the leverage can be increased by having a narrow and very much longer cage, although this possesses the disadvantage of ends with corners in which animals tend to come to rest as a result of their thigmotactic reactions.

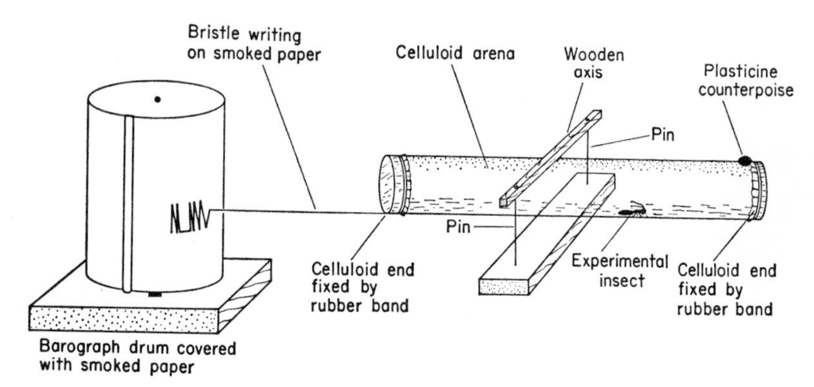

Fig. 4–7 Micro-aktograph apparatus.

Instead of an electrically driven smoked drum, a clockwork barograph drum can be employed as a miniature kymograph. This also serves to reduce the overall size of the apparatus so that it can be placed in an incubator or refrigerator, and the light and temperature controlled artificially. From a purely ecological point of view, however, it is usually not necessary to delve too deeply into the physiology of the biological clock.

If the animal you are investigating is too small to tip any micro-aktograph, you will have either to work for months constructing some complex photo-electric aktograph or else to spend a few sleepless days and nights in visual observation. Actually this need not be as exacting as it sounds. Suppose you decide to use the percentage of animals crawling on the surface of a box of leaf-litter as a criterion of activity—you will probably need to make counts at intervals of 3 hours throughout the day and night, and so you can get some rest in between. Moreover, you will probably find that a peak of activity occurs in the early hours of the evening with, perhaps a minor peak at dawn. In this case, you may be able to reduce the number of midnight and daytime counts.

4·5 Seasonal changes

Seasonal changes in the fauna of microhabitats provide a fertile field for investigation, both from the point of view of synecology and of autecology.

Seasonal rhythms are apparent in the lives of almost all animals and I will mention only a few here. Many small invertebrates have a life-cycle of a year's duration, but some cryptozoic forms such as centipedes and millipedes may live for much longer. In either case the timing of the seasonal rhythm may vary in different species so that inter-specific competition is avoided. For example, in Germany, the wolf-spider *Lycosa amentata* matures during March and April, *L. pullata* reaches a peak in May, whilst the majority of specimens of *L. tarsalis* do not undergo their final moult until June or July. Again the greatest number of *Coelotes inermis*, *Zelotes latreillei* and *Agroeca brunnea* reach maturity in the spring with a very much smaller peak in autumn, whilst the congeneric *C. atropos*, *Z. pratensis* and *A. proxima* show a major autumnal peak, many fewer reaching maturity in spring (TRETZEL, 1956).

Seasonal changes are apparent too in the behavioural responses of many species. Thus, a marked seasonal change occurs in the humidity responses of the millipede *Schizophyllum sabulosum* in Finland. The dry summer reaction is gradually reversed to moist in the autumn when the animals hibernate. This seasonal change correlates well with the ecology of the species: whereas in summer it is often to be found in dry places, it always hibernates in moist surroundings (PERTTUNEN, 1953). A reversal of the normal humidity response also occurs in females about to lay eggs. These show a positive reaction to moisture which takes them to the damp localities where oviposition occurs.

Seasonal changes have also been observed in the humidity reaction of the common earwig, *Forficula auricularia*. In summer, the animals show a clear and strong 'preference' for the drier side of a humidity choice-chamber apparatus (section 4.3) but, in winter, they respond positively to moist air which causes them to burrow into the soil for hibernation. Various species of beetle have also been found to show seasonal changes in behaviour, but there is much need for further work of this nature (section 6.3).

The duration of daylight may play a great part in controlling seasonal rhythms in arthropods. It may also initiate *diapause*, a dormant state that enables many organisms to persist in inconstant environments and regions that would otherwise be unfavourable for permanent habitation. Diapause is under hormonal control. It is usually characterized by reduced metabolism and enhanced resistance to climatic factors such as heat, cold or drought. In the case of *phytophagous* or plant-eating insects, the onset of diapause frequently coincides with some distinctive phase in the growth cycle of the host plant, yet experiments have revealed that there may be no causal connection between these events. Diapause is usually triggered by changes in photo-period but it is independent of light intensity provided that this exceeds a threshold value greater than that of moonlight.

Synthesis 5

At the roots of marram grass growing in sand dunes in Europe, one may sometimes find a species of false-scorpion, *Dactylochelifer latreillei*, in abundance. *D. latreillei* is a rare species and occurs only in a few localities even though it may be quite numerous in these. I have kept a specimen for months in a corked tube, feeding it on clothes-moths, mosquitoes and other small insects. The species is resistant to desiccation, catholic in diet and apparently not restricted to its peculiar habitat by characteristics of the physical environment. What then confines *D. latreillei* to sand dunes? We do not know the answer to this question but I suspect that it may be that in most places *D. latreillei* is somehow less efficient than other species of false-scorpion, and is therefore eliminated by *competition*. In the harsh environment of the sand dunes, however, many of these others are unable to exist and those that can may be handicapped even more than *D. latreillei*. This suggests that a species does not necessarily exist in the habitat most favourable for it, but in that in which it competes most satisfactorily with others.

5.1 Populations

Both experimentally and in nature, three primary factors may cause a fall in the reproductive rate of a population of animals and thereby regulate its number. These are: exhaustion of the food supply, adverse 'conditioning' of the medium with excretory products, and increased density. The first two seldom occur naturally and certainly not with the inhabitants of soil, leaf-litter, and other microhabitants, with the possible exception of certain specialized predators. Furthermore, it is most unlikely that density operates directly; although soil animals and litter animals may reach astronomical numbers, their populations are not exceptionally dense in view of their small size.

From an ecological point of view, an animal does not exist independently of its environment which includes its rivals, predators and parasites, as well as physical factors. At the present time there is considerable disagreement amongst zoologists as to the relative importance of these factors. According to one school of thought, the agents which determine population densities in nature can be divided into two groups: density-dependent and density-independent. Density-independent factors are defined as those which bear equally heavily upon a population, whether its density be great or small: the most drastic of these are physical and climatic. But, it is argued, if a particular factor of the environment is to *regulate* the population density of a species by holding it within definite limits, it must be able to destroy a greater fraction of the population of that species when its density is high than when it is low. If destruction were merely proportional to population density, the factor causing it could not be regulatory. Hence climate alone cannot control or regulate animal numbers.

For example, if a hard winter killed, say, 99 per cent of a particular species of insect, the survivors might be able, during the next season, to increase to a far larger number than before. And even if 99 per cent of these were eliminated during the succeeding winter, there would still be many more survivors than there had been the previous year, and so on. Of the density-dependent mortality factors which play most heavily upon a population when it has reached maximum density, competition between individuals for food and shelter, predators, disease and the effects of migration are probably the most important.

Some zoologists believe, however, that natural control is effected chiefly by climate and that it is unnecessary to invoke density-dependent factors to explain either the maximum or minimum numbers occurring in a natural population. Certainly many insect species seem to spend most of the time recovering by annual increases from occasional climatic set-backs. No doubt the truth lies somewhere between these extreme views. Possibly climate may control the more normal fluctuations in the number of a species while density-dependent factors are significantly involved only when maximum or minimum densities are approached. In any case, it is clear that population sizes are directly controlled by environmental influences of one sort or another. If these are sufficiently complex, no one species can become too numerous: but where they are simpler, it is easier for plague proportions to be reached. The fauna of leaf-litter, with its richness of species, is a comparatively stable one and any unusual change in a particular species tends to be buffered by the remainder of the population.

5.2 Food chains

Regular sampling shows that both diurnal and seasonal rhythms in numbers, activity and distribution of animals occur. For the significance of these to be appreciated it is necessary first to investigate the food chains involved. Thus, vast numbers of spring-tails in the top soil may provide food for smaller numbers of predatory insects such as ground beetles and their larvae, spiders, centipedes and so on which, in turn, may be eaten by even smaller numbers of larger animals such as frogs, lizards or birds. Thus in a single community there is a pyramid of numbers whose basal layer is the abundant mass of vegetable matter such as fallen leaves, while the herbivores and carnivores constitute successive layers of rapidly decreasing size, until there is a small number of larger predators at the apex.

The construction of food chains and analysis of the pyramid of numbers is an essential preliminary to an understanding of the biological structure of any community of animals. The various species should be assessed as to numbers, stage of development, food, predators, parasites and so on.

Information about food preferences and other aspects of behaviour can be obtained by experiments with captive animals or by dissection of recently collected specimens, whilst the presence of young or larval stages and their proportions to the numbers of adults at different times of the year may give some idea about life-cycles and mortality rates.

5·3 Summary

Mere collecting and the assimilation of raw data is not enough: the object of this Study is to encourage the student to find out something new for himself. There are so many problems regarding the fauna of microhabitats that can profitably be investigated that he will have almost unlimited scope. The first essential is to select a problem that can be expressed simply in a few words. For example, how is the fauna of leaf-litter affected by the types of leaves or kind of soil on which they are resting? What is the significance of pH, drainage, shadow, shelter from the wind or direction and slope of the ground? How do the numbers and proportions of the different species fluctuate throughout the seasons or from year to year? Does the fauna of a particular type of microhabitat vary in different parts of the country and is this related to climate or to some other factor? Which of the species show diurnal rhythms of activity; do they come to the surface at night and burrow into the soil during the hours of daylight? Do they climb trees? Is such movement affected by temperature, moisture and winds? For, in addition to problems of ecology and distribution, various aspects of the behaviour of the fauna are well worth attention.

A comprehensive answer to quite simple questions such as these can often be obtained only by a great deal of laborious research which extends the question far beyond its original scope. If you begin with a complex problem, the chances are that you will not get anywhere! It is better to begin with a simple autecological problem and let this lead on later to more ambitious synecological projects.

A Specific Example 6

Not everyone can get away whenever he pleases to study ecology in the country. But, even in the most densely populated cities, there are parks and gardens which provide a variety of microhabitats of the kind that I have been describing. Their value as subjects for instruction is therefore high. Consider, for example, a rubbish heap or a rotting log with loose bark under which live numbers of cryptozoic animals. If several students are taking part in the investigation of a single microhabitat it is a good plan for each to select a particular group or species of animals for study. Later, all this autecological data can be assembled and a picture of the synecology of the microhabitat constructed (section 6.4). In this chapter I shall concentrate on woodlice, as their biology is particularly well known.

First, it will be necessary to find out what species are present. The five

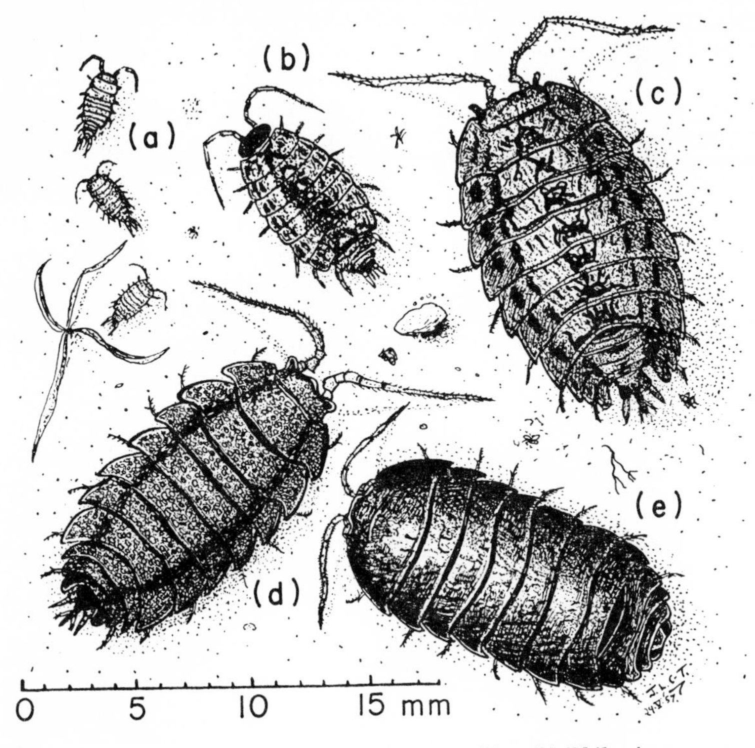

Fig. 6–1 British woodlice: (**a**) *Trichoniscus pusillus*; (**b**) *Philoscia muscorum*; (**c**) *Omiscus asellus*; (**d**) *Porcellio scaber*; (**e**) *Armadillidium vulgare.* (Reproduced from *Country-side*, **18**, 105, the Journal of the British Naturalists' Association)

most common British woodlice, illustrated in Fig. 6–1, are easily recognized by the naked eye. It is quite possible that all of them will be present in the same microhabitat. On the other hand, aggregations of a single species are by no means uncommon even when several species are abundant locally. Of these facts there is, at present, no satisfactory explanation. Using the methods outlined in Chapter 2, it can be shown that the temperature is fairly constant and the relative humidity of the air within the chosen habitat is very high, both at night and during the day. (If the air were not almost saturated few, if any, cryptozoic animals would be living there, as can be demonstrated by comparison with other microhabitats where the air is less humid.) The soil pH should also be measured (section 2.7): woodlice are usually more plentiful in alkaline than in acid habitats although most species can live in both.

6.1 Water relations

All woodlouse species tend to lose water comparatively rapidly in dry air (Fig. 3.1). The most resistant in this respect in Britain is the pill woodlouse, *Armadillidium vulgare* which can survive in dry places for several hours and is often to be seen running about in the open during the daytime. Next in the series comes the common *Porcellio scaber*, easily recognized by its rough, granular integument; then the garden slater *Oniscus asellus* and finally the smaller *Philoscia muscorum* which soon dies from desiccation when exposed to dry conditions (Fig. 6–1). All species must spend the greater part of their time in an atmosphere that is saturated with water vapour, but there is considerable variation regarding their ability to withstand dry air, high temperatures and the period of time during which they can venture into dry places (EDNEY, 1954) (section 6.2). All this is assumed from the results of physiological experiments: it has not yet been fully confirmed by observations in the field although some information has been obtained about migration up trees (BOER, 1961; BRERETON, 1957).

Air humidity fluctuates between wide limits, except in microhabitats where a high moisture content is usually retained. However, since the normal excretion of woodlice cannot entirely compensate for water-uptake in nearly saturated air, a long stay in such conditions becomes unfavourable and the animals must from time to time come out into the open and lose excess water by transpiration. Hence they tend to climb trees, walls and buildings at night.

Woodlice are able to extract water from their food. In this way, they can make up part of the water lost by evaporation. Of the three species that have been investigated in this respect, *A. vulgare* is the most efficient, *P. scaber* comes next and *O. asellus* last. It might have been expected that *O. asellus* which, of the three, loses water most rapidly by transpiration, would have made up for it by a greater ability to extract water from food. But this is not the case. Adaptation to life on land does not concern a number of isolated characteristics of animals, but their organization as a whole (KUENEN, 1959).

Despite the restrictions inherent in the physiology of their water and temperature relations, woodlice are by no means unsuccessful terrestrial animals. PARIS (1963) has argued that Isopoda invaded the land, despite their physiological handicaps, because of the advantage of a readily accessible trophic niche afforded by dead vegetation. The ubiquity and success of *A. vulgare* may be due to the fact that it can readily change its diet from grazing to detritus feeding (PARIS and SIKORA, 1965).

6.2 Diurnal rhythm

Woodlice are nocturnal in habit and, apart from *A. vulgare*, do not normally wander abroad much until after dark. As it is always dark within their microhabitat, they would not know when to come out if it were not for their endogenous rhythms or biological clock (section 4.4). By means of choice-chamber apparatus (section 4.3) it has been found that the positive response to humidity of *Oniscus asellus* is less strong in darkness than in light, and still less in the nocturnal phase: it increases with desiccation. Movement away from light is enhanced in animals which have been in darkness for some time, but desiccated animals tend to become photo-positive.

These experimental results can be related to the nocturnal ecology of the species. Woodlice often wander in dry places at night and this is permitted by the reduction in the intensity of their humidity responses. The increased photonegative response after they have been conditioned to darkness ensures that they get under cover promptly at daybreak and thus avoid many potential predators such as birds. On the other hand, if their daytime habitat should dry up, woodlice are not restrained there until they die from desiccation since they tend to become photopositive in dry air and thus are able to wander in the light until they find some other damp hiding place and again become photonegative (CLOUDSLEY-THOMPSON, 1952).

When different species of woodlouse are compared, it is found that the degree of nocturnal activity is correlated with the ability to withstand water-loss by transpiration. *Philoscia muscorum*, for example, transpires more rapidly, is more strongly photonegative and more strictly nocturnal than *Oniscus asellus* or *Porcellio scaber*, whilst *Armadillidium vulgare* is the least intensely nocturnal and photonegative (CLOUDSLEY-THOMPSON, 1956). This behaviour is part of the osmoregulatory system of the animals whose nightly excursions are related in frequency, duration and direction by their water balance and the humidity of the air. As we have seen (section 6.1), water is taken up during the day faster than it can be excreted and this is evaporated at night (BOER, 1961). Finally, the nocturnal emergence of woodlice is inhibited by wind because air currents tend to remove the shell of moist air that surrounds that transpiring animal.

From this example, it will be seen that, in order properly to understand the ecology of an animal, it is necessary to study its reactions both during the day and at night. This does not mean, however, that the student has to work every night in the field or laboratory. Diurnal rhythms can be reversed experimentally by keeping animals in darkness during the day time and illuminating them at night with an electric light controlled by a time switch. After a couple of weeks in such conditions, the experimental animals will be in the nocturnal phase during the hours of daylight and only a few night-time checks with controls are required to confirm that their reactions during the actual day are the same as those at night of control animals kept in natural daylight and darkness.

6.3 Seasonal rhythms

We now know that the daily life of woodlice is regulated by the interaction of their biological clocks, humidity and light responses. I have already mentioned seasonal changes in the responses of millipedes to humidity (section 4.5). The same occur in woodlice which show a marked rise in the intensity of their humidity responses in the spring, when the rains bring them out of hibernation (GUPTA, 1963). At the same time, seasonal changes occur in the distribution of woodlice. Whereas *Phil. muscorum* remains under stones and litter throughout the year, the tiny *Trichoniscus pusillus* moves in summer from its winter habitat under stones to litter and dead wood. *O. asellus* is found mainly under stones throughout the year but in summer it also occurs on dead wood and trees. Finally, *P. scaber*, whose winter habitat is at the base of trees, tends to move upwards during summer (BRERETON, 1957). The physiological significance of this has already been mentioned (section 6.1).

In addition to these seasonal changes, the log under which the woodlice are living will become more and more rotten as time passes so that the investigators will have to be alert to spot changes that are due to succession (section 1.4) and distinguish them from the effects of other environmental factors.

6.4 Population studies

Having related the physical features of the immediate environment of the woodlouse to its physiological responses and capacities, we can see how beautifully it is adapted to its habitat. This is only a beginning, however. The next step is to investigate its food, predators and parasites—indeed, all its interactions with other organisms. From this, we must then pass on to consider populations of woodlice and of other animals living in the same microhabitat, using the methods outlined by JACKSON and RAW (1966). Finally a diagram should be constructed of the pyramid of numbers and food-webs in the microhabitat, showing quantitatively the interactions between the various species present. Only then will a synthesis have been achieved between the autecology of the various species and the synecology of the microhabitat. If you get this far, you will see that the distinction between autecology and synecology is superficial and that both are part of an ecological whole. It is this 'whole' that one must try to understand.

Further Reading

ALLEE, W. C., EMERSON, A. E., PARK, O., PARK, T. and SCHMIDT, R. P. (1949). *Principles of Animal Ecology*. W. B. Saunders, Philadelphia & London.

CARTHY, J. D. (1958). *An Introduction to the Behaviour of Invertebrates*. George Allen & Unwin, London.

CARTHY, J. D. (1965). *The Behaviour of Arthropods*. Oliver & Boyd, Edinburgh & London.

CARTHY, J. D. (1966). *The Study of Behaviour. Studies in Biology No. 3*. Edward Arnold, London.

CHAPMAN, R. N. (1931). *Animal Ecology with Especial Reference to Insects*. McGraw-Hill, New York & Maidenhead.

CLOUDSLEY-THOMPSON, J. L. (1968). *Spiders, Scorpions, Centipedes and Mites*. 2nd edn. Pergamon Press, Oxford.

CLOUDSLEY-THOMPSON, J. L. (1960). *Animal Behaviour*. Oliver & Boyd, Edinburgh & London.

CLOUDSLEY-THOMPSON, J. L. (1961). *Rhythmic Activity in Animal Physiology and Behaviour*. Academic Press, New York & London.

CLOUDSLEY-THOMPSON, J. L. (1965). *Desert Life*. Pergamon Press, Oxford.

CLOUDSLEY-THOMPSON, J. L. (1965). *Animal Conflict and Adaptation*. G. T. Foulis, London.

CLOUDSLEY-THOMPSON, J. L. (1972). *The water and temperature relations of woodlice (Isopoda: Oniscoidea)*. Merrow, Watford.

CLOUDSLEY-THOMPSON, J. L. (1969). *The Zoology of Tropical Africa*. Weidenfeld and Nicolson, London.

CLOUDSLEY-THOMPSON, J. L. and CHADWICK, M. J. (1964). *Life in Deserts*. G. T. Foulis, London.

CLOUDSLEY-THOMPSON, J. L. and SANKEY, J. (1961). *Land Invertebrates*. Methuen, London.

DOWDESWELL, W. H. (1952). *Animal Ecology*. Methuen, London.

DOWDESWELL, W. H. (1959). *Practical Animal Ecology*. Methuen, London.

EDNEY, E. B. (1957). *The Water Relations of Terrestrial Arthropods*. The University Press, Cambridge.

FRAENKEL, G. S. and GUNN, D. J. (1961). *The Orientation of Animals*, 2nd edn. Dover Publications, New York.

FRANKLIN, T. B. (1955). *Climates in Miniature*. Faber, London.

GEIGER, R. (1952). *The Climate near the Ground*. Harvard University Press, Cambridge, Mass.

HARKER, J. E. (1964). *The Physiology of Diurnal Rhythms*. The University Press, Cambridge.

JACKSON, R. M. and RAW, F. (1966). *Life in the Soil. Studies in Biology No. 2*. Edward Arnold, London.

KALMUS, H. (1948). *Simple Experiments with Insects*. Heinemann, London.

KEVAN, D. K. MC. E., ed. (1955). *Soil Zoology*. Butterworths, London.

KEVAN, D. K. MC. E. (1962). *Soil Animals*. H. F. & G. Witherby, London.

KUHNELT, W. (1961). *Soil Biology*. (Trans. N. Walker.) Faber, London.

LAWRENCE, R. F. (1953). *The Biology of the Cryptic Fauna of Forests*. A. A. Balkema, Cape Town.

MACFADYAN, A. (1963). *Animal Ecology, Aims and Methods*, 2nd edn. Pitman, London.

MURPHY, P. W., ed. (1962). *Progress in Soil Zoology*. Butterworths, London.

PEARSE, A. S. (1939). *Animal Ecology*, 2nd edn. McGraw-Hill, New York & Maidenhead.

RUSSELL, E. J. (1957). *The World of the Soil*. Collins New Naturalist, London.

SANKEY, J. (1958). *A Guide to Field Biology*. Longmans, Green, London.

SAVORY, T. H. (1955). *The World of Small Animals*. University of London Press, London.

References

ASCHOFF, J. (1960). *Cold Spring Harb. Symp. quant. Biol.*, **25**, 11–28.
BOER, P. J. DEN (1961). *Arch. Néerl. Zool.*, **14**, 283–409.
BOYOUCOS, B. A. and MICK, A. H. (1940). *Tech. Bull. Mich. (St. Coll.) agric. Exp. Stn.*, **172**, 1–12.
BRADY, C. (1965). *Sch. Sci. Rev.*, **46**, 724–633.
BRERETON, J. LE G. (1957). *Oikos*, **8**, 85–106.
CLOUDSLEY-THOMPSON, J. L. (1952). *J. exp. Biol.*, **28**, 165–172.
CLOUDSLEY-THOMPSON, J. L. (1956). *J. exp. Biol.*, **33**, 576–582.
CLOUDSLEY-THOMPSON, J. L. (1962). *A. Rev. Ent.*, **7**, 199–222.
DENTON, R. J. (1951). *Bull. Am. met. Soc.*, **32**, 214–216.
EDNEY, E. B. (1953). *Bull. ent. Res.*, **44**, 333–342.
EDNEY, E. B. (1954). *Biol. Rev.*, **29**, 195–219.
GHILAROV, M. S. (1958). *Proc. X Int. Congr. Ent.*, **2**, 725–730.
GREAVES, T. (1964). *Aust. J. Zool.*, **12**, 250–262.
GUPTA, M. (1963). *Proc. natn. Inst. Sci. India* (B), **29**, 203–206.
KEMPSON, D. A. and MACFADYAN, A. (1954). *J. anim. Ecol.*, **23**, 376–380.
KOIE, M. E. (1954). *Oikos*, **4**, 178–186.
KROGH, A. (1940). *Ecology*, **21**, 275–279.
KUENEN, D. J. (1959). *Entomologia exp. appl.*, **2**, 287–294.
PARIS, O. H. (1963). The ecology of *Armadillidium vulgare* (Isopoda: Oniscoidea) in California grasslands: food, enemies, and weather. *Ecol. Monogr.*, **33**, 1–22.
PARIS, O. H. and SIKORA, A. (1965). Radiotracer demonstration of isopod herbivory. *Ecology*, **45**, 729–734.
PENMAN, H. L. and LONG, I. (1949). *J. Scient. Instrum.*, **26**, 77–78.
PERTTUNEN, V. (1853). *Ann. Soc. zool.-bot. fenn. Vanamo*, **16**, 1–69.
SCHNEIRLA, T. C., BROWN, R. Z. and BROWN, F. C. (1954). *Ecol. Monogr.*, **24**, 269–296.
SOLOMON, M. E. (1957). *Bull. ent. Res.*, **48**, 489–506.
TRETZEL, E. (1956). *Z. Morph. Ökol. Tiere*, **44**, 43–162.